Sustainable cities

HC779

Urban Studies

Sustainable cities: Japanese perspectives on physical and social structures

Edited by Hidenori Tamagawa

United Nations University Press

TOKYO · NEW YORK · PARIS

United Nations University Press
United Nations University, 53-70, Jingumae 5-chome,
Shibuya-ku, Tokyo, 150-8925, Japan
Tel: +81-3-3499-2811   Fax: +81-3-3406-7345
E-mail: sales@hq.unu.edu general enquiries: press@hq.unu.edu
http://www.unu.edu

United Nations University Office at the United Nations, New York
2 United Nations Plaza, Room DC2-2062, New York, NY 10017, USA
Tel: +1-212-963-6387　Fax: +1-212-371-9454
E-mail: unuona@ony.unu.edu

United Nations University Press is the publishing division of the United Nations
University.

Cover design by Mea Rhee

Printed in India

ISBN 92-808-1124-X

Library of Congress Cataloging-in-Publication Data

Sustainable cities : Japanese perspectives on physical and social structures /
edited by Hidenori Tamagawa.
    p. cm.
 Includes bibliographical references and index.
 ISBN 928081124X (pbk.)
 1. City planning—Japan. 2. Sustainable development—Japan. I. Tamagawa,
Hidenori, 1956–
HT169.J3S87   2006
307.1′2160952—dc22   2006003758

Contents

List of tables and figures

Figures

Contributors

Peter John Marcotullio
Fellow
Institute of Advanced Studies at
    United Nations University
Nishi-ku
Yokohama 220-0012
Japan
pjmarco@ias.unu.edu

Yoshitsugu Aoki
Professor
Department of Architecture and
    Building Engineering at Tokyo
    Institute of Technology
Meguro-ku
Tokyo 152-8552
Japan
aoki@arch.titech.ac.jp

Makoto Maruyama
Professor
Graduate School of Arts and Sciences
    at University of Tokyo
Meguro-ku
Tokyo 153-8902
Japan
maruyama@waka.c.u-tokyo.ac.jp

Tokue Shibata
Professor Emeritus at Tokyo Keizai
    University
Former Director of the Department of
    General Planning and Coordination
    at Tokyo Metropolitan Government
3-7-17 Amanuma
Suginami-ku
Tokyo 167-0032 (home)
Japan

Takashi Kawanaka
Head
Urban Development Division at the
    Urban Planning Department,
    National Institute for Land and
    Infrastructure Management
Tsukuba
Ibaraki
305-0802 Japan
kawanaka-t92j5@nilim.go.jp

Kiyoshi Takami
Department of Urban Engineering at
    University of Tokyo
Bunkyo-ku

Tokyo
113-8656
Japan
takami@ut.t.u-tokyo.ac.jp

Hidenori Tamagawa
Professor
Graduate School of Urban Science at
    Tokyo Metropolitan University
Hachioji
Tokyo 192-0397
Japan
htama@comp.metro-u.ac.jp

Nobuhiro Ehara
Staff
Urban Development Headquarters at
    Tokyu Corporation
Shibuya-ku
Tokyo 150-8511
Japan
nobuhiro.ehara@tkk.tokyu.co.jp

Junko Ueno
Doctoral Candidate
Department of Sociology at Sophia
    University
Chiyoda-ku Tokyo
102-8554

Japan
j-ueno@sophia.ac.jp

Masahisa Sonobe
Professor
Department of Sociology at Sophia
    University
Chiyoda-ku
Tokyo 102-8554
Japan
m-sonobe@sophia.ac.jp

Yoko Shimizu
Lecturer
Graduate School of Allied Health
    Sciences at Tokyo Medical and
    Dental University
Bunkyo-ku
Tokyo 113-8510
Japan
shimizuyo.chn@tmd.ac.jp

Osamu Soda
Professor
School of Social Sciences at Waseda
    University
Sinjuku-ku
Tokyo 169-0051
Japan
sohda@waseda.jp

Acknowledgements

This book is based on nearly 50 workshops and numerous discussions held over four years in relation to the Joint Research Project II "Comprehensive Studies on Recycling-oriented Society and City Planning" conducted by the Centre for Urban Studies of the Tokyo Metropolitan University (represented by Professor Shunji Fukuoka in FY1998 and FY1999, and by Professor Hidenori Tamagawa in FY2000 and FY2001). We regret that we could only include a small proportion of the studies in this book, but we would like to thank everyone involved in the joint project for their contribution and cooperation. The chapters written by Japanese authors were based on the translation of the 23rd volume of *Urban Studies Series* (*Toshi Kenkyu Sosho* in Japanese) at Tokyo Metropolitan University, and are included by permission of the Centre for Urban Studies. We thank the Centre for its generosity.

We would also like to thank Mr. Scott McQuade, Publications Officer, and Ms Yoko Kojima, Senior Publications Coordinator, of the United Nations University Press, and Mr. Nobuyuki Kawade of the United Nations University Institute of Advanced Studies for their dedicated efforts which have made the publication of the translated book possible.

Lastly, the translation, proofreading and publication of this book were funded by a Grant-in-Aid for Publication of Scientific Research Results from the Japan Society for the Promotion of Science (JSPS). The publication of this English version would not have been possible without this grant, and we would like to express our sincere gratitude to the JSPS for all their assistance.

Foreword: Geographical background

Japan is an archipelago consisting of five major islands (Honshu, Hok-kaido, Shikoku, Kyushu and Okinawa) and thousands of smaller ones. The entire land area is about 380,000 km^2 (about one twenty-fifth of that of the US). The total population amounts to 125,000,000 (almost half that of the US). Figure 0.1 shows the location of the country. The plains, chiefly on the coasts, are the most urbanized regions in the world, while 75 per cent of the hinterland is preserved as forest. About 80 per cent of the population resides in cities.

Japan has experienced rapid industrialization and urbanization since the late nineteenth century. She lacks most natural resources and has to import them for industrial use. Chapter 3 of this book should be understood in this context.

The Tokyo Metropolis and the Tokyo Metropolitan Region, located in central Honshu, is the most populated region in this country. The urban problems of Japan are apparent there. Figure 0.2 shows Tokyo and the surrounding prefectures. Railways criss-cross the region like blood vessels. Tokyo also has a radial and ring road network. Chapter 4 describes the long-term garbage and land problems of Tokyo Metropolis over the nineteenth and twentieth centuries.

The population of Tokyo metropolis is 12 million, while that of the To-kyo Metropolitan Region (Tokyo Metropolis and prefectures of Chiba, Saitama and Kanagawa) is around 34 million. The analyses of Chapter 7 concern the Tokyo area, which means the Tokyo Metropolitan Region and includes the southern part of Ibaraki prefecture.

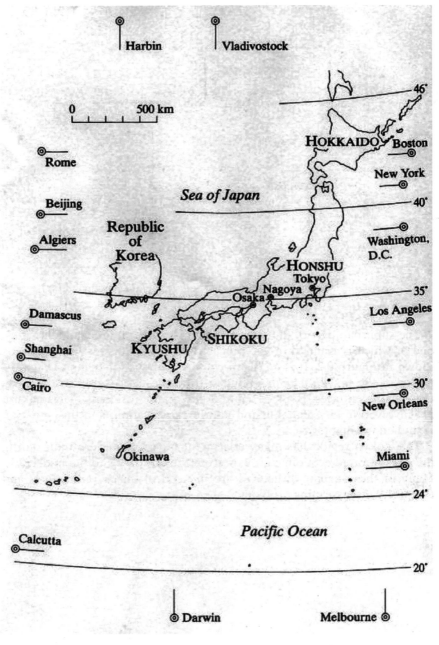

Figure 0.1 Map of Japan (*Source*: Shibata, 1993)

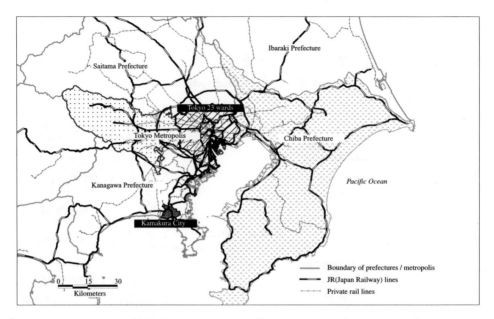

Figure 0.2 Map of the Tokyo area (*Source*: created by the editor)

The suburban part of the Tokyo region has several independent cities, one of which is Kamakura City, located 40 km south-west of Tokyo, as shown in Figure 0.2. It is an historical city, famous for being the seat of power of the first *shogunate* of Japan from the late twelfth to the early fourteenth centuries. Being a cultural city, many residents of Kamakura are involved in city planning and nature preservation, and this situation is studied in Chapter 8.

The Tokyo region has many apartment blocks that were built during the days of rapid economic growth, from the late 1950s to the mid-1970s. Many of these ageing structures are in need of major renovation, and Chapter 9 discusses this problem.

Abbreviations

General:

COD	chemical oxygen demand
ICLEI	International Council for Local Environmental Initiatives
NPO	non-profit organization
SCAP	Supreme Commander for the Allied Forces
TSP	total suspended particles
UNCHS	United Nations Centre for Human Settlements (Habitat)
WCED	World Commission on the Environment and Development
WWF	Worldwide Fund for Nature (formerly known as the World Wildlife Fund)

Japan:

DIDs	densely inhabited districts
DP	population change (per cent)
JR	Japan Railway Company
TDM	transportation demand management
TMG	Tokyo Metropolitan Government

State of Oregon:

LCDC	Land Conservation and Development Commission
OARs	Oregon Administrative Rules

SDCs   System Development Charges
TODs   Transit Oriented Developments
TPRs   Transportation Planning Rules
TSPs   State Transportation System Plans
UGB   Urban Growth Boundary
VMT   vehicle miles travelled

UK:

DPs   development plans
LTPs   local transport plans
PPGs   Planning Policy Guidances
RPBs   Regional Planning Bodies
RPGs   Regional Planning Guidances
SPs   structure plans

Introduction

Hidenori Tamagawa

"Sustainability" is a popular word nowadays, frequently heard in the media. It is often used in the context of global environmental issues – for example, at the environmental summits and in various environmental appeals. This book, which is an anthology of articles on urban sustainability, deals with the concept of "sustainability" from slightly different dimensions and aspects to those usually applied.

The meaning of "sustainable" here is similar to that used in the term "sustainable community", but we aim to make it broader in concept, clearer in image, and deeper in insight, as outlined below. The authors have attempted to expand and redefine the concept of "sustainable" from various perspectives of engineering, humanities, social sciences and mental health.

As this may be the first attempt to introduce the Japanese discussion on sustainability to other nations, we hope it may trigger further discourse on the topic. The term "sustainability" is complex and still not clearly defined. At a global level, for example, the term is usually considered in the context of natural environmental protection, population growth and the economic gaps between advanced and developing countries. Here, we shall illustrate how the term is used in Japan from the following two aspects: (a) man-made systems in Japan have traditionally managed to coexist with the natural environment, but due to rapid urbanization in recent years, these systems have become damaged, and we now need to restructure them using new technologies and methods;

1

(b) there are yet many challenges we need to address for the well-being of citizens and to make the "urban society" sustainable.

The topics discussed in the book may sound less pressing and problematic than the many serious environmental and population issues existing in the developing countries, and one might argue that since damaged man-made systems can be restored and improved simply through technology transfer, they do not warrant international discussion. We believe, however, that the urban problems now confronting Japan and its people are not unique to that country: they will emerge when any country reaches a certain level of development. Successful technology transfer is not necessarily achieved through the simple application of technology, but is often hampered by personal, social and cultural barriers within the developing nation. We believe that the experiences of the Japanese, who have undergone an unprecedented level of rapid economic and urban growth, will surely provide precedents for urban researchers and practitioners worldwide, and will facilitate further international exchanges.

Japan first experienced industrial revolution a century ago and has since developed a flourishing modern technology, including information technology (IT). Thus the speed of modernization for Japan has been more rapid than in most of the Western world, although somewhat slower than that experienced in many of the currently developing countries. As for Japanese cities, many were destroyed or badly damaged during World War II and were subsequently reconstructed in a temporary or somewhat makeshift fashion. Thus the rapid postwar growth of Japanese urban areas was similar to that currently seen in developing countries. For example, the pattern of construction of new towns followed the Western method of developing splendid modern residential areas. However, in view of the current declining birthrate and the growing proportion of elderly people, as in the West, these ageing neighbourhoods are now urgently in need of reconstruction.

Hence, modern Japanese cities have characteristics of both those of the West and those of the developing countries. Thus the case studies and perspectives of Japanese cities discussed in this book can also be applicable to other countries that are in similar stages of development.

Rather than provide an in-depth approach to a specific field of urban studies, this book takes a cross-sectional approach by experts across a variety of academic fields and illustrates both the diversity and the shared values held by urban scholars in approaching the concept of urban sustainability.

Shared values here might be summarized as the twin notions that (a) the system is self-sufficient in principle and yet is open to the external world, and (b) the system stimulates society to be free from the illusions of eternal expansion and growth. Such a system should be positive and ready to change in our daily lives while not sacrificing future generations. In fact we might change the spatial catchphrase "Think globally, act locally" to "Think eternally, act temporarily". Though it is a difficult challenge, the book contains many issues and ideas for solutions to achieve such a system.

The spatial systems of buildings, apartment complexes, urban districts, small cities, metropolises, metropolitan regions, local regions and nations are all self-sufficient to some extent. At a time when concern for the sustainability of cities in difficult economic times is required, we need to generate a lively discussion on the level and type of self-sufficiency and openness of these systems, or, in other words, how they can be structured "physically" and "socially".

Each chapter of the book consists of an independent article or editorial with introductory remarks to enable a smooth transition from one chapter to the next. The book starts with a global and general discussion of the subject, followed by consideration of conceptual approaches to sustainable systems and urban sustainability together with historical outlines of metropolitan problems. This is followed by several case studies of both technological and sociological issues in current Japanese metropolitan areas. The book concludes with an outline of a strategy for a smart, or enhanced, community.

The first chapter attempts to distinguish between problems of developing world cities and those of the West, focusing on a comparison of the experiences of cities of the Asia-Pacific region and those in the United States. Starting from the viewpoint which associates long waves of development with the Western experience, this chapter demonstrates that the current urban development context is significantly different. Specifically, it is argued that the new development context has telescoped the aspects of "urban environmental transition" (McGranahan, Jacobi, Songsore, Surjadi and Kjellen, 2001). Globalization, demographic shifts, advanced technologies and institutional and political change have led to the "time–space telescoping" of urban environmental transitions such that the urban challenges faced by cities in the past over a longer period and in a serial manner now occur more rapidly and in an increasingly overlapping manner. The implications are significant in that previous solutions to urban problems, which appeared slowly and were implemented under specific institutional contexts, are no longer appropriate.

The second chapter serves as a bridge between the discussions of the first chapter on urban sustainability from a global viewpoint and discussions in later chapters on urban sustainability in Japan. Following the rapid industrialization of the last century, Japan now finds itself in a post–industrial age with many consequent physical and sociological changes within its cities. The sustainability of cities – the major human stage for both business and residential activities in society – is now an important issue. The focusing on Japanese cities, which are positioned at the interstice of oriental complexity and Western order, is particularly instructive for both the developed and the developing regions. Keeping in mind this systematic explanation of observed differences from a global viewpoint, we gradually move on to focus on Japanese problems and proposals.

First, however, Chapter 2 considers the general concept of "sustainability", which serves as a theoretical basis for general sustainable systems. An attempt is made to illustrate clear images of sustainability, "dynamic balance", "stable balance" and "probabilistic equilibrium" by the use of many easy-to-understand metaphors and in a way that demonstrates how "sustainability" can be applicable to diverse subjects. The author, who has specialized in the analysis of urban space, then concludes the chapter by focusing on sustainability of urban land use. A simple conceptual model, such as that employed here, can provide a useful reference frame for the consideration of urban areas as dynamic systems. Similar models are introduced in Chapters 7 and 10.

Chapters 2 and 3 describe the general outline and historical perspectives of Japanese metropolitan environmental problems, and provide a helpful introduction to the Japanese situation for readers who may not familiar with it.

Chapter 3 also discusses the issue of sustainability from a broad economic perspective within the grand framework of civilization. The author refers to the issue raised by Karl Polanyi and Yoshiro Tamanoi – namely, that the word "economy" conjures up for us of two different spheres – and discusses two steps to make the economy (in real and practical terms) sustainable: firstly, by linking agriculture and industry based on organic production; and, secondly, by creating an autonomous and regional-based group or a community between the community economies at local level and the global economy. From this point of view, the author proposes a society whose regional wealth consists not only of properties and stocks but also of the social capital which enriches people's freedom.

The author of Chapter 4, Tokue Shibata, has been involved for many years in environmental administration in the Tokyo Metropolitan Government. Beginning with a discussion of the garbage disposal problems that Tokyo confronted in its early days of urbanization, he cites numerous cases where "land" is at the root of Tokyo's environmental problems. The author shares his wealth of accumulated experience with readers to provide insights into the future "recycle-oriented and sustainable city of Tokyo".

Subsequent chapters focus on specific, current cases in Japan. Chapter 5 looks at the sustainability of cities from the perspective of energy-conscious buildings and urban planning. The author, Takashi Kawanaka, has experience in various projects related to this theme, and proposes a set of formulae for energy-conscious urban planning. He follows this with reviews of the current Japanese legislative urban planning system from the viewpoint of "formulae" applications.

Control of automobile use is a significant issue in saving energy in modern cities, and Chapter 6 examines this issue in relation to a retail development control system. The author, Kyoshi Takami, is a young researcher of transportation planning, and he gives a detailed introduction of two cases abroad, drawing on policy issues to be addressed by Japan.

Chapter 7 discusses urban sustainability from the view of population stability. The authors, Nobuhiro Ehara and Hidenori Tamagawa, conducted a quantitative study in the Tokyo Metropolitan Region using National Census data from 1970 to 1995, and identified 25 districts where the population had remained stable. By analysing the population age structure of these districts the authors were able to determine that the trends affecting young people seem to be the key to the population stabilization of these districts.

In the next two chapters the focus is on urban citizens, not as "mass" but as major players in urban life. Chapter 8 discusses citizens' environmental protection movements in Kamakura City in the medium- to long-term perspective. Its authors, Junko Ueno and Masahisa Sonobe, are both sociologists and recognize that citizens' movements aimed at the protection of the environment are an important way in which ordinary people can become involved in the governmental decision-making processes. They compare and contrast cases of "anti-development movements" and "environmental protection movements" and, in this highly motivated analysis, illustrate differences, such as "sustainability of the movement itself", and similarities, such as the concept of "joint right of possession".

Chapter 9 focuses on the "emotional" aspect of urban citizens. Through a survey of the rebuilding of public apartment complexes, the author, Yoko Shimizu, who is an expert of community health care, examines how a changing living environment influences residents' lives, and how personal networks and exchanges among neighbours are maintained or altered. This important study is likely to attract considerable attention in this field, since the number of apartment complexes built during the high-growth period now in need of renovation or replacement is increasing nationwide.

Finally, Chapter 10 looks again at the issue of sustainability from a general perspective. Based on the assumption that a sustainable city requires a radical change of local communities, the author, Osamu Soda, demonstrates images of the "shape", "mechanism" and "resources" of sustainable communities, and examines the principles of the driving forces leading to these images. At the centre of the discussion are the ecological concepts referred to in Chapters 2 and 7. The author interprets and organizes these concepts from a social scientist's perspective, and concludes with an outline of the "smart human community".

We hope that this book will give readers an opportunity to learn about some of the Japanese experiences in this field and set them on a journey to explore further the sustainability of cities: a journey that has just begun.

1

Comparison of urban environmental transitions in North America and Asia Pacific

Peter John Marcotullio

Introduction

The environmental problems currently being experienced by developing cities are significantly different from those experienced by Western cities during the nineteenth and early twentieth centuries. This notion has been bolstered by research on globalization, urban development and urban environmental conditions (see for example, Burgess et al. 1997). Studies demonstrate, for example, an association between the urban environment and the function of the city in the regional and global economy (Lo and Marcotullio 2001). This chapter furthers these arguments. It attempts to differentiate the development contexts and therefore the experience of environmental risks between the now developed and developing world. These differences are reflected in variations between urban environmental transitions, and are most easily seen in comparing Western urban history with that of rapidly growing cities of the Asia Pacific. The determinants of these changing patterns include a time- and space-related effect called *time-space telescoping*: the shifting in the timing, speed and sequencing of previously experienced development patterns so that they occur sooner, faster and more simultaneously (Marcotullio 2005). This chapter concentrates on identifying the more simultaneous aspect of *time-space telescoping*.

The chapter is divided into four more sections. The next and second section discusses the theories used to analyse the differences between the developed world experience and those of the developing world. This section includes a brief overview of the factors contributing to transitions amongst sets of environmental concerns and how various phases in Western urban history have been associated with different types of environmental challenges. Also presented is a comparison between the Western model and the contemporary "telescoped" model. It also elaborates on the mechanisms of *time-space telescoping*, a perspective derived from the work of McGranahan and his colleagues (2001) on urban environmental transitions.

The third section examines the developed world experience of urban environmental transitions, using the USA as an example.[1] Specific environmental conditions are associated with various long waves of development. The attempt is to associate specific environmental burdens to socio-techno-economic systems. Transitions between environmental challenges are defined by technological, economic and institutional change, prompted in many cases by political crises resulting from overwhelming environmental burdens. Also described is the changing nature of theoretical understanding of environmental problems and how these advances have impacted upon the social, technical and political relationship to urban environmental conditions.

The fourth section presents a set of analyses that demonstrate the simultaneous nature of urban environmental challenge experiences in the current developing world. It begins with an analysis of the global distribution of environmental burdens. This is followed by an outline of how the experiences of cities in rapidly developing Asia are different from those in the now developed world through an analysis of UN-habitat urban data. The fifth section concludes the chapter.

Conceptualizing the underlying reasons for differences

The environmental burdens experienced by urban citizens at any given time are the result of a number of factors. While the full examination of all factors is beyond the scope of this paper, several important influences can be identified and analysed. The history of environmental conditions in any city is associated with economic activities, particularly when externalities of production bring negative consequences. The environment is also impacted by the types of technologies of the era as, while technologies bring benefits, they also have unintended outputs that often produce environmental challenges. Another important influence on the urban environment is the total societal understanding of what the problems are and how to solve them. Hazardous environmental conditions may not be understood or even identified. The "solutions" to environmental problems have also undergone their own history, which is intertwined with changing institutions and the politics over aspects of the built environment. Finally, the environmental conditions within and around cities are related to the natural pre-conditions and the history of the use of the ecosystems in and around cities by societies. This section attempts briefly to put these factors into a perspective that allows for the comparison of urban environmental transitions across time and cultures. The perspective starts by placing the influence of these various factors within the theory of long waves of development. It then uses the notion of urban environmental transition to identify the types of environmental concerns associated with each era. Further, it outlines how these factors could be seen as different in rapidly developing Asia Pacific cities as compared to how they were experienced by the now developed world. These differences are a result of what is called the *time–space telescoping* of transitions. Embedded in this notion is the fact that current experience of urban environmental challenges in rapidly developing world cities has been "telescoped", creating overlapping sets of challenges.

Factors impacting urban environmental change

The theory of long waves of economic development was proposed by Nikolai D. Kondratiff (1979). Using available data on Germany, France, England and the United States of America, he demonstrated, among a group of variables,[2] a secular trend in a specific direction that was structurally linked to the overall changes in the economic environment of the particular society. These waves, of 40–60 years in length, are characterized by accelerating rates of price increases from deflationary depression to inflationary peaks, followed by decade-long plunges from the peaks to primary toughs, which again are followed by weak recovery, and then by stages into the next deflationary depressions. There is debate over the underlying reasons why these exist (see, for example, Maddison 1991: 112)[3] but the general consensus is that long-term fluctuations in national economies of some fifty years duration have occurred.

Associated with the long-term trends in price cycles are shifts in technologies. Brian Berry (1997), for example, suggests that US history is marked by a rise and fall of a succession of "techno-economic systems", defined by interrelated sets of technologies with which are associated sets of raw materials, sources of energy and infrastructure networks. The first set of techno-economic systems is associated with the use of wind, water and wood. This was followed by the coal, steam and iron system, and then steel, kerosene and electricity. In the fourth phase the technologies underpinning growth included petroleum, internal combustion and chemicals. The fifth way, which we may be experiencing now, may be driven by business services, new telecommunication and biotechnologies and a host of other new innovations.

History is not driven by advances in technology alone, however. Technological and economic change, moreover, do not explain how environmental conditions shift within societies. Often social crises and political action motivate change that leads to transitions from one set of problems to another. Political actions take place within an institutional framework that also has shifted over time and hence has impacted the character of environmental and urban politics.

Related to political struggles over the environment is the theoretical understanding of what problems are and how they could be solved. Our understanding of health and the environment has changed dramatically over the last 200 years. This understanding of the human-natural environment, however, has often been the result of the study of unintended consequences of previous actions. Moreover, while advances have been made, they have often been ignored in favour of conditions suitable to the most powerful in societies.

None of these factors in and of themselves explain the changes in urban environments, but together they make a powerful collection of forces that help to outline the urban history of the developed world. Yeates (1998), for example, has used many of these factors to define four periods to study changes in urban patterns. In his schema, which loosely follows long waves of development patterns, these eras include frontier mercantile (to 1845), early industrial capitalist (1845–1895), national industrial capitalist (1895–1945), and mature capitalist (1945–present). Onto these different eras of development can be mapped shifts in urban environmental challenges and for most of those in the US, their "solutions". "Solutions", in this sense, often meant the export, both geographically and into time (i.e. the future), of the particular set of environmental problems. At the source of these problems, however, for that time and those places, conditions improved.[4] Table 1.1 attempts to summarize the environmental challenges faced by cities in the US during the different eras associated with techno-economic long wave cycles. The details of this table will be explored in the next section.

Urban environmental transitions: comparing historical experiences between the developed and developing world

In a recent text that sums up 10 years of research in this field, Gordon McGranahan and his collaborators (2001) present a persuasive argument concerning the relationship between development, affluence and the urban environment. They claim that urban environmental burdens tend

Table 1.1 Techno-economic systems, urban development eras and associated urban environmental challenges

Kondratieff Cycle	Technology systems	Urban phase	Environmental challenges
To 1845 A-Phase (peak 1814) B-Phase (trough 1843)	Wind, water, wood	Mercantile cities	Major health impacts: small pox, yellow fever, convulsions, cholera; Infrastructure demands: water supply, street refuse, street drainage; Other conditions: domestic animal waste
From 1845 to 1895 A-Phase (peak 1864) B-Phase (trough 1896)	Coal, steam and iron	Early industrial cities	Major health impacts: pneumonia, tuberculosis, diarrhoea; Infrastructure demands: housing, transportation, water supply and sewerage; Other conditions: high densities, inadequate health care systems, horses;
From 1895 to 1945 A-Phase (peak 1920) B-Phase (trough 1930s)	Steel, electricity and kerosene	National industrial cities	Major health impacts: pneumonia, tuberculosis, diarrhoea, occupational; Infrastructure demands: paved roads, water, sanitation and treatment, solid waste; Other conditions: Soot/air pollution, rising occupational hazards
From 1945–Present A-Phase (peak 1970s) B-Phase (?)	Petroleum, internal combustion engines and new chemicals	Mature/post-industrial cities	Major health impacts: circulatory system, cancer, nephritis, pneumonia including influenza, new diseases (HIV/AIDS, SARs, etc); Infrastructure demands: wastewater treatment, incineration water filtration, watershed protection, auto transport, other energy; Other conditions: air and water pollution, cumulative persistant chemical and toxic waste (including nuclear waste), non-point source pollution, consumption-related issues (land use, energy, natural capital, etc), global concerns (ozone depleting and GHGs emissions), emerging pollutants (PM2.5)

Sources:
Kondratieff cycles: Kondratieff (1979); Yeates (1998: 64) (citing Mager 1987).
Technology systems: Berry (1997: 302); Lo (1994).
Urban phases: Yeates (1998); Melosi (2000).
Environmental challenges: Health-related from Jackson (1985); Water supply, sewerage and sanitation from Melosi (2000) and Chudacoff (1981); other from Melosi (2000), Hays (1987), and various sources.

to be ecosystem threatening, dispersed and delayed in impact in affluent cities, while in less wealthy cities environmental risks are health threatening, localized and immediate in impact. Importantly, the model is based upon the notion of increasing scale of impact that comes with wealth. Accordingly, urban activities impact larger areas with growing wealth. As a poor city moves beyond the "brown" agenda, for example, environmental impacts of cities increase in scale from the household and neighbourhood levels to city-wide regions. For those cities, struggling with the "green" agenda, the dominant environmental impacts of urban-based activities are regional if not global (e.g. greenhouse gas and ozone depleting substance emissions). The brown agenda prioritizes immediate conditions: environmental health and local issues relating to inadequate water and sanitation, indoor air quality and solid waste disposal. The green agenda, on the other hand, focuses on ecological sustainability and addressing issues related to resource degradation, contributions to global environmental burdens and other extra-urban problems.

It is possible to use this theory in combination with the factors described above to compare the speed and sequencing of the emergence of environmental conditions between the now developed world and the rapidly developing world in the Asia Pacific. The question the chapter addresses is: do urban environmental transitions hold similarly for both cities that developed earlier as well as those undergoing rapid development now? It is hypothesized that there are important differences between the two experiences, one being the telescoping of the timing between transitions (Figure 1.1). As cities are now developing under strong pressures of globalization, they are influenced by a variety of different demographic and technological shifts, and lack the institutions and capacity to deal with challenges related to these new pressures; they are experiencing sets of environmental problems in a different fashion than were experienced by the now developed world.

The current era of development is driven by the globalization process (Dicken 1998; Held et al. 1999; Johnston et al. 1995; Knox and Agnew 1998). Globalization can be defined as a deepening, thickening and speeding up of economic, social and political interdependencies. As globalization and technological breakthroughs have lowered prices and brought goods to many nations, particularly in Asia Pacific, so it has changed the dynamics of the urban environmental transition. Increasingly, coastal cities along an axis from Tokyo to Jakarta are becoming centres of industry. This intensive and rapid development has been facilitated by global flows of trade and investment that are larger than ever previously experienced. Moreover, it has come with particularly intensive environmental challenges (Asian Development Bank 1997).

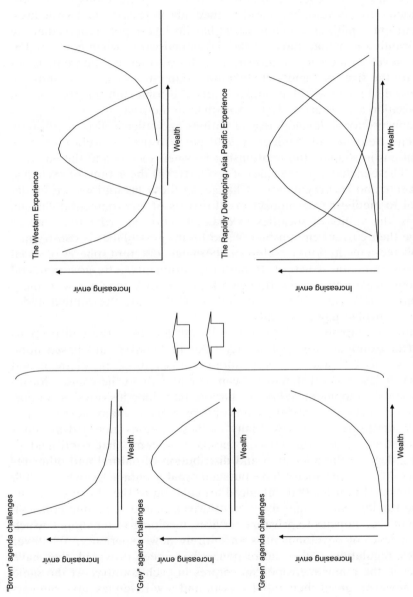

Figure 1.1 Schematic difference between Western and current experience of the urban environmental transition (*Source*: Marcotullio and Lee, 2003)

13

In addition, demographic shifts have allowed concentrations of lower income groups to be incorporated into upper and middle income groups within these cities with large differences in access to these same goods and services. The important difference between populations in the now developed and developing world is their absolute sizes and structures. The current rapidly developing world has far larger populations that are significantly older than those of the developed world when compared at similar levels of income. Moreover, the roles of cities are different, as not only are the sizes of rapidly developing Asian cities larger than those of the now developed world at similar levels of income, but the processes of urbanization are occurring faster now than in the past.

The institutional development of nations and cities is also significantly different. Bennett and Dahlberg (1990) point to the importance of three institutions in shaping the relationship between society and the environment. These include: (i) the rules of property; (ii) the actions of exchange (markets); and (iii) regulations. Changes in these institutions are fundamental to mediating the impact of all drivers of environmental change. That is, the ability of societies to channel action in order to create or change their current circumstances and future possibilities is constrained by how these institutions develop and permeate different spheres of social life. Governmental structures (types of governments, intra-governmental relations, legal and judicial frameworks, etc.) work together with these institutions to create political possibilities (i.e. create the context under which political struggles unfold).

Within the rapidly developing Asia Pacific countries, these institutions vary (for example, private property rules in Malaysia are based upon those of the UK and are very different from those in Vietnam and China). A few general statements can be made about the context for institutional development. While the nations have long histories as entities, they have emerged as modern states only recently, as many were colonies of now developed countries. Many of these nations emerged as states with centralized governments or one-party systems that controlled to varying degrees the allocation and distribution of natural and other resources. Most governments have pursued rapid economic growth through governmental policies that enhanced articulation to the global economic system (under export- and investment-driven macro-economic policies). These include policies to privatize services, including water supply, which at their level of development is an entirely new experience. Whatever policies, regulations, rules and responses have been provided, they have privileged the major metropolitan centres in each country. At the same time, however, given their rapid growth and low revenues, governments have not been able to supply infrastructure at rates comparable to the now developed countries at similar levels of income.[5] As a result, these institutional differences between Western countries and the Asia–Pacific region impact on the numbers and types of political actors and the distribution of influence among them, and therefore impact on the issues that get onto the local political agenda and the priorities among those that do.

Because of these local conditions, the nations and cities within the Asia Pacific have a unique relationship to Kondratieff cycles. They are the latecomers of the fourth wave and, as such, have not fully adopted the accumulation regime that is now experienced by the already developed world. It is only a short time since these developing nations were largely traditional agricultural systems without industry, and as they become articulated to the global economic system they experience both time- and space-related effects that impact upon the relationships between, *inter alia*, development and urban environment conditions. Time-related effects include technological improvements and associated decreases in environmental impacts per unit consumed; space-related effects include processes that concentrate activities in geographically uneven patterns and in increasingly concentrated spaces. This concentration may be of infrastructure, economic activity, people, technologies, and so forth. Globalization has space-related effects by, for example, concentrating certain types of infrastructure – communication, transportation, financial and business services, headquarters of transnational companies – in specific locations such as major world cities (Friedmann 1986; Friedmann and Wolff 1982; Lo and Yeung 1998; Sassen 1994), while new transportation and communications technologies have transformed the ability of information, goods and services to move across space (Dicken 1998).

Combining these two effects suggests that activities within lower and middle-income cities articulated to the global city network will be mediated by an increasingly diverse set of technologies that includes both the latest advances together with outdated but previously often used ones, and that these will be adopted by progressively more diverse populations – including traditional populations and higher income groups inhabiting a more global living space – all with a greater variety of environmental conditions within which to navigate. The ramifications of these effects on society as a whole are already apparent in the collapsing, telescoping and compression of the sequential development patterns of the past. Consequently, environmental changes are now emerging more rapidly and taking effect simultaneously across regions (Marcotullio 2005). This development pattern is termed *time–space telescoping*, and this study will highlight the simultaneous aspects occurring with the emergence of these environmental changes.

The Western experience

Environmental and urban historians in the US have segmented urban history in a number of ways (Chudacoff 1981; Hays 1987; Melosi 2000; Monkkonen 1988; Yeates 1998). The periods chosen often correspond loosely to those defined by long waves of development, possibly because many of these histories emphasize the role of economic activities and technology changes. In line with the argument presented, this section highlights the economic and technical underpinnings of US urban development patterns along with crisis issues that have helped to stimulate political change and therefore the shift in environmental agendas. In doing so, the shifting nature of growing wealth and institutional development is associated with various types of environmental conditions. At the same time, this section also attempts to illustrate the serial nature of these problems by demonstrating the length of time between their various critical points. The section is divided into four sub-sections: (1) mercantile cities (pre-1845) and local environmental challenges; (2) early industrial cities (1845–1895), the increasing intensity of local environmental challenges and the emergence of regional environmental challenges; (3) advanced industrial cities (1895–1945), the increasing regional challenges and the emergence of non-bacteriological challenges; and (4) the emergence of post-industrial cities (1945–present), industrial pollution and larger ecosystem challenges.

Mercantile cities (pre-1845) and local environmental challenges

During the mercantile era, wealth was created through trade and the accumulation of tangible assets such as land and gold. The US economy was based primarily on agriculture and the export of staple products, and most cities were centres for the import of manufactured products from Europe and the export to Europe of fish, furs, timber and agricultural goods. Consequently, the largest urban areas were located along the Atlantic coast, although significant cities grew along major waterways.

Economic, and hence urban, growth was slow. The paucity of data allows only best guesses, but scholars suggest that between 1607 and 1776 economic growth for colonial cities was between 0.3 and 0.6 per cent per year (Bruchey 1992). By the time of the War of Independence (1776–1783), the population of the colonies was approximately 2.3 million, only a small proportion of whom lived in cities. In 1790, there were five cities with 10,000 or more inhabitants, and by 1830 this number had risen to 23.

These towns grew slowly. For example, during the entire period Baltimore had fewer than 800 housing starts per year (Yeates 1998).[6] Outward growth of cities was limited by transportation technology and animate-power (largely related to horses, carriages and walking). The internal structure of these cities reflected their economic and social nature, energy and the technological limits of the time. Shipbuilding and repair yards, places of business of the financial institutions, merchants and shopkeepers were all clustered close together and around waterfronts. The urban fabric was mixed use, with residents and commercial industries huddled close together in a dense fabric for safety, convenience and social necessity. Many cities also had walls surrounding their cores.

Urban environmental problems were local in nature. Urban populations were small enough to allow for solutions that avoided confronting the dilemma of defining the public interest and allowed for the exploitation of ecosystem services within the city boundaries. The solutions, however, had mixed results. Early settlements were not sanitary places, particularly for the poor. High human densities, mud and stagnant water, farm animals, refuse tossed into streets, conflagrations and epidemic disease outbreaks were common. At first smallpox was the most frequent and dreaded disease, but by the early 1800s yellow fever was ravaging cities, particularly in the Northeast (Chudacoff 1981). At that time, the urban populace had only vague notions of the underlying causes of problems such as the spread of diseases, and the lack of technological development in terms of adequate health services and water supplies meant there were no satisfactory solutions available to them. Social and political issues were largely settled by elites.

One of the first solutions to these urban challenges was street grading and the provision of side gutter systems (first implemented in Boston in 1713), although interest in street paving would not reach a peak until the end of the century. The initial reason for these developments was to facilitate the removal of horse waste in order to provide a safe and healthy gathering space for the local populace (McShane 1999). Epidemics and fires were fought through the introduction of prototype water supply systems in the largest cities and were largely in the hands of the private sector. These systems were crude and limited. The Manhattan water supply system, a privately owned experiment by Aaron Burr, was one of the most notorious of this period.[7] At the same time, sanitation, in terms of both wastewater and solid waste removal and management, continued to be the responsibility of individuals (Melosi 2000) and was largely insoluble, leaving citizens simply to adjust to the conditions. Towards the end of this period, a powerful scientific notion, the "sanitary idea", emanating from England, linked filth with disease and provoked a rationale and strategy for improving conditions in cities. Under this theory, "miasmas" (filth and foul smells) were believed to be responsible for epidemics. While not in fact the root cause, the theory placed emphasis on environmental sanitation, which became the prominent aspect from then on and throughout the next environmental period.

Early industrial cities (1845–1895), increasing intensity of local urban environmental challenges and the beginnings of regional environmental challenges

During the early periods of industrialization in the US, manufacturing and processing plants using capital initially derived from trade, or through inflows of investment from Europe, created profits from the sale of manufactured goods. Most of the manufacturing firms were small and locally owned and their growth was nurtured behind protectionist trade barriers and high transportation costs. New York City, for example, had thousands of small industries creating a variety of goods. Economic and urban growth increased at more rapid rates. Between 1820 and 1870, growth rates in the USA increased to 1.3 per cent per annum (Crafts 2000).[8]

Industrialization became synonymous with urbanization. The number of urban centres increased from 939 to 2,262 between 1880 and 1910, and the number of cities with populations over 100,000 increased from 19 to 50. By 1880, approximately 14 million people lived in cities (Yeates 1998). Within the interior of the country, cities also began to appear, helped in part by the construction of canals and improved transportation to the interior.[9]

Immigration, which had been spurred on by events such as the Irish famine, increased both productivity and densities within the cities. By the end of this period, New York City (which at the time was coterminous with the New York City boundary) was reaching its highest population size and density.[10] Many of those living in the city were crammed into dense low-income neighbourhoods and small rooms without ventilation or sunlight (Riis 1972).

The development of transportation technologies, such as steam engines in trains and ships, allowed cities to extend their borders outward. This, combined with aggressive real estate expansion and border adjustment through governmental annexation or consolidation, dramatically influenced the size of cities. Chicago, for example, increased its area in 1888 through annexation from 36 to 170 square miles. Greater New York was formed in 1897 out of Manhattan, Brooklyn, Long Island City, Richmond County, most of Queens, some of Westchester, and parts of Kings County.[11] Interestingly, rather than keeping cities compact, it was the provision of public transportation to areas beyond the city boundaries and the single fare (nickel train ride) that facilitated the early stages of the suburbanization and decentralization processes (Jackson 1985).

With rapid growth and the beginnings of industrialization, the various urban environmental problems that had emerged during the previous period grew to critical levels, setting off a number of crises. The first and foremost of these problems was access to drinkable water. Providing water became an increasingly important part of urban public services, particularly after diseases such as cholera and yellow fever affected the elites (Miller 2000). The provision of an adequate water supply became the first important public utility in the United States and the first municipal service that demonstrated a city's commitment to growth. City officials and urban booster interests, such as chambers of commerce, boards of trade, and commercial clubs, promoted a variety of downtown improvements, which often included providing a healthy community as an essential ingredient. Moreover, the growing civic consciousness was promoted by dramatic crises during the period, such as the Civil War Draft Riots, which in part was a response of the immigrant community to terrible housing conditions.[12]

The increasing understanding of the causes of disease further enhanced the strength of those economically minded urban boosters. In the latter half of the nineteenth century, faith in environmental sanitation as the primary weapon against disease lost followers as the era of bacteriology emerged. Sanitary reforms since the 1840s had emphasized the prevention of disease through the clean-up of space and increasing moral fibre. With advances in the medical community and the emergence of bacteriology, emphasis was placed on finding cures for disease, and this brought a sea change in the way health and disease were viewed.[13] The germ theory – or bacteriological theory – of disease was well established as a scientific fact by the 1880s through the work of Louis Pasteur, Robert Koch, and others. While the old view of public health was concerned with the environment, the new perspective of public health was concerned with the individual (Melosi 2000).

These forces, when taken together, increasingly promoted centralized organizational structures and capital-intensive technical innovations for water supply and, later, wastewater removal and treatment.[14] Early in this period, only the largest cities had reasonably efficient water systems.[15] By the late nineteenth century all communities needed city-wide waterworks. In 1870, the total number of waterworks stood at 244; by 1890 that number had tripled to 1,879. In 1880, 64 per cent of cities had waterworks; in 1890, approximately 72 per cent had them. Indeed, the number of waterworks in operation increased more rapidly than did the population in the decade between 1880 and 1890 (Melosi 2000).

In contrast to the growth of water supply systems, underground wastewater systems were meagre and only began to appear later. As one historian noted, this was because the problem of water removal only emerged after water supplies started to bring in large amounts of water from outside the city (Tarr 1996). Further, few cities were in a position to finance these two major technologies of sanitation simultaneously. Some private investments were obtained for early development of water supply systems, but a similar path for sewerage systems was unlikely, leaving all in the hands of the public sector. Therefore, privy vaults and cesspools were relatively widespread. As water supplies continued to provide only the necessary levels for hygienic life, these technologies and open ditches, as storm drains, were common.

This situation only began to change with the implementation of "piped-in" water and the "water closet". These facilities began to appear in some middle-class homes soon after running water supplies became more readily available. In 1865, New York City had only about 10,000 water closets for a population of 630,000; Boston with a population of 180,000 had only 14,000 in 1864 and Buffalo with a population of 125,000 had just 3,000 (Melosi 2000). It was only after these technologies were implemented in higher income neighbourhoods that underground sewers were recognized as valuable in staving off epidemics and in offering other benefits.

The solution of better sanitation systems for the removal of wastewater, however, led to further problems. Joel Tarr (1999) suggests that ironically, as sewer systems became more successful in capturing wastewater, they become greater pollution threats. This is particularly true for cities that employed combined systems without wastewater treatment. The most common, albeit the easiest and cheapest, form of sewage disposal practised in the United States was dilution as American cities relied on their vast water resources to serve as sinks for many years. In 1900, approximately 90 per cent of the nation's sewage was dumped into water bodies without treatment. Some cities – Chicago, for instance – were using the same water body for both drinking and dilution purposes. Other downstream cities were subject to drinking water increasingly contaminated by upstream users.

As the solution to the growing wastewater problems facilitated their spilling over to larger geographical areas, some conditions inside the cities improved. Technical improvements in water supply and wastewater disposal led to greater water demand and, to some degree, of freedom from epidemic diseases. At the same time, growing populations increased the demand for water, further increasing the waste load on rivers and streams. Economic and demographic forces were also creating increased pollution loads on many recipient rivers. River pollution, however, was almost completely ignored unless impacting on powerful political forces downstream. The politics of this type of pollution favoured the upstream city, as there were no regulations to prevent this type of pollution and the actions helped to provide better internal environments. Politicians, decision-makers, elites and concerned citizens turned a blind eye to the results. Thus, the period ended with the unintended result of causing high typhoid death rates in downstream cities that drew their water supplies from rivers into which upstream cities discharged their sewage.

Water supply and wastewater removal remained the main focus of urban environmental specialists. Refuse collection, on the other hand, was not well developed at this time. Street cleaning was largely confined to wealthy and commercial areas. While, in New York City the first permanent incinerator was built on Governor's Island, it was only when a reform government took hold and George Waring was placed as Street-sweeping Commissioner in 1895 that public systems began to address the host of environmental issues surrounding solid waste.

Air pollution problems centred on the burning of coal, horse manure on the streets and the foul smell emanating from refuse. Citizens were asked to endure these difficulties as solid waste management was expensive and coal smoke from industries was equated with growth. Since providing water and better health care had increased life expectancy to 46 years for men and 48 years for women these issues were either ignored or not considered priorities. In the next phase, as disease deaths became disassociated from urban life, more attention was paid to wastewater, solid waste and the new growing threat of industrial pollution.

National industrialization (1895–1945), increasing regional environmental challenges and the emergence of non-bacteriological challenges

At the turn of the century, the US was entering a period of increasing industrial might. Powerful national and international corporations were emerging and large-scale assembly line manufacturing produced a vast array of consumer items. During this period, the US embraced Fordism – the introduction of mass production – and the mass consumption society emerged.

Growth was stimulated by a set of technological advances discovered at the end of the previous era, including the harnessing of electric power for lighting (1882) and for vehicles (1886), the development of steel alloy (1891), the telephone (1876) and the internal combustion engine (1883). In some cities, particularly those of the Eastern Seaboard, it was further enhanced through a second and larger wave of immigrants (approximately 25 million) that entered the country between 1880 to 1920.[16] Most of these people came from Southern and Eastern Europe.

This era began with a great urban political consolidation wave. In the middle of this period, by 1920, the nation changed from a rural to urban one as the numbers of residents in towns and cities escalated. Urbanization continued its rapid pace throughout the era, and between 1920 and 1940 the urban population of the US increased from 54.2 million to 74.4 million, and the number of urban areas with populations of 2,500 or more grew from 2,722 to 3,464. In 1920, 51.2 per cent of the country was urbanized; by 1950, the figure had reached 64 per cent.

As metropolitan areas grew in population they increased in land area, and densities continued to decrease around the cores. Although deconcentration of the largest metropolitan areas began before 1900, large-scale changes were profound in the twentieth century. First, by 1902 over 50 per cent of streets in the largest cities and 16 per cent of all city streets were paved with asphalt. By 1924, almost all cities had all their streets paved. Metropolitanism, the geographic configuration of urban areas centred on a consumer society based upon car travel, increasingly became the defining characteristic of the American landscape. Such reorganization drew urban and rural areas closer together and made them more interdependent. Indeed, changes in cities impacted the character of rural hinterlands, which then impacted the preferences and tastes of urban citizens (see, for example, Cronon 1991).

Urban factories were responsible for 90 per cent of the country's industrial output. It was the cities of the Northeast and along the Great Lakes where the most dynamic development took place. By 1920, nine of the ten largest cities in the country were in this manufacturing belt. Maturing industrial facilities centred in cities were matched by new commercial and service activities reflecting the transformation of the American economy after World War I from a primarily industry-based economy to a consumer-based economy.

The urban environmental challenges experienced by cities in the nineteenth century changed in both type and extent. Foremost on the list were water quality, wastewater removal and treatment and increasing solid waste generation. Also during this period the industrial pollution, while not a major concern, began to impact waterways.

Water supply solutions provided cities with the ability to continue growing and, indeed, to grow to large sizes. Increasing social concern over water purity, particularly by cities downstream of larger cities, combined with advances in microscopic studies and filtration technologies that facilitated both the characterization of bacteria and the elimination of the worst water-borne diseases. With the discovery of typhoid bacillus, a scourge until the 1920s, and other pathogenic microorganisms, public health officials and engineers began to understand the extent of water-borne diseases, and sought methods for combating them. Attention focused primarily on filtering and treating water. Advances in these areas helped to break the association of density and disease.[17]

Bleaching powder, chlorine gas and the use of chloride of lime as treatment to water enhanced purity. With these techniques a dramatic decline in typhoid fever rates followed. By 1920 typhoid death rates in the United States were more in line with those of European countries, such as England, Germany and France, where water purification had been widely applied. Aggregate typhoid (and paratyphoid) fever death rates per 100,000 persons dropped steadily. Between 1920 and 1945, the rate dramatically dropped from 33.8 to 3.7 (Melosi 2000).[18] By 1940, the urban mortality penalty in the US had been eliminated; it became healthier to live in American cities than to live in rural areas (Haines 2001).

Despite the Great Depression of the 1930s, national trends in the construction and expansion of waterworks continued to grow steadily due to the infusion of federal funding of the New Deal. In 1924, there were approximately 9,850 waterworks in the US; in 1932, there were 10,789, in 1939, 12,760 and approximately 14,500 in 1940. Although the rate of growth was strongest from the 1880s through to the early 1920s, increases in the 1930s were also significant (Melosi 2000).

Disposal of wastewater, due to massive increases in "piped-in" water, continued to be a stimulus to underground sewerage development. By 1920, 87 per cent of the urban population lived in sewered communities, which represented about 45 per cent of the US population. These figures were impressive because the number of sewered communities had increased from approximately 100 in 1870 to 3,000 in 1920. The growth of sewered urban population continued to rise steadily through to the 1940s and reached almost universal status by the end of World War II, at least in the largest cities, extending to 8,917 communities (Melosi 2000).

It was not until the 1930s that the increase in population living in communities having treatment other than dilution caught up with the overall growth of the population in sewered communities. Given the success of filtered water, investments in these infrastructures increased. Between 1920 and 1945 the population served by sewage-treatment plants increased from 9.5 million to 46.9 million, or by 494 per cent. By 1945, 62.7 per cent of people lived in sewered communities with treated sewage facilities, with the remaining 37.3 per cent living in communities that still disposed of raw sewage.

By comparison to water supply and wastewater removal provisions, disposal and collection of refuse continued to remain a local issue. Refuse was a problem that appeared to be site-specific as opposed to having a larger geographic impact. With the demise of the Filth Theory of disease, refuse disposal (until later in the twentieth century) declined in significance as a way to avert severe health hazards.

A substantial effort towards organized public systems of collection and disposal did begin around 1895. Accumulating household wastes, ashes, horse droppings, street sweepings and rubbish facilitated awareness of a growing "garbage problem". Higher health consciousness was helped as the US shifted from a producer society to a consumer society and the quantities and varieties of wastes increased rapidly. Indeed, the waste stream became a testament to growth as well as to increasing productivity and the consumption of goods (Melosi 2000).

Slowly, American cities began to implement systematic refuse collection and disposal systems. Two disposal methods dominated solid waste management processes: dumping refuse on land or into water, an immediate solution that merely shifted the problem from one site to another and using refuse for farms (including fertilizer and animal feed). For example, by 1927, 27 per cent of the garbage collected in 27 major cities was openly dumped, 20 per cent of all garbage collected from 21 cities was used on farms (Melosi 1999). Within cities, especially those not situated on waterways, it was common to dump refuse on vacant lots or near the "least desirable" neighbourhoods. Dumping into water was practised by New York City, which in 1886 unloaded 1,049,885 of its 1,301,180 cartloads of refuse into the ocean and for many years continued to rely on this practice as a primary means of disposal.

Few people gave much thought to the relationship between rising affluence, the consumption of goods and the increasing production of waste. Faith in science and technology to overcome mounting refuse levels infused throughout the general population. The incinerator and the automobile were two technologies that people thought would relieve the municipal waste load. As a primary mode of transportation, cars brought the hope that city streets would be free of manure and other debris (Tarr 1993), while incinerators could reduce the volume of trash by simply burning it. Unfortunately, the ironic result was that horse manure and solid waste would be supplanted by greater environmental dangers in the form of noxious fumes.

Massive quantities of refuse continued to mount in the cities of the interwar years. While the devastation of the Great Depression reversed the escalating economic growth of the 1920s, it only temporally derailed the consumer trends and habits begun in the 1890s and elaborated throughout the early twentieth century. The generation of solid wastes rose steadily between 1920 and 1940 from 2.7 pounds per capita per day to 3.1 pounds per capita per day. Street sweepings continued to rise even with the phasing out of the urban horse (Melosi 2000).

Late in this period, new sets of pollutants were emerging in air and water ecosystems. Industrial wastes and other toxic materials started to grow. Not until after World War I were industrial wastes viewed as a problem that affected water purity. These threats started to mount as the nation continued to industrialize. In 1923, a report indicated that in no fewer than 248 water supplies had been affected by industrial wastes throughout the US and Canada (Melosi 2000).[19] These conditions and the shifting priorities of cities set the stage for the next era.

The emergence of post-industrial cities (1945–present): industrial pollution and larger ecosystem challenges

Under advanced Fordism, the US combined mass production of goods with mass consumption to produce widespread material growth and decreasing social inequities. This form of accumulation continued until the middle of the 1970s. Thereafter, the most recent period of American history is associated with the current manifestation of globalization processes (see, for example, Schaeffer 1997) and the emergence of a post-Fordist society; characterized by flexible forms of production, growing inequity and slower growth. Indeed, the post-war history for the world can be sub-divided into two phases: the "Golden Age" and the "Landslide" (Hobsbawn 1996). During the Golden Age, directly after World War II when the US emerged as the most powerful economic and military country in the world, its economy experienced rapid and sustained growth. From 1949 to 1973, per capita wealth (in constant dollars) increased by over two-thirds as the national GDP per capital increased annually on average by 2.4 per cent (Crafts 2000). One result was a shift in the centralization and concentration of economic might to large corporations. For example, in 1973 the 500 largest industrial corporations accounted for 80 per cent of all corporate profits in the US and together employed more than 15.5 million individuals.

At the same time the population increased rapidly (by more than 60 million in two decades) – a process that created the baby boom, which peaked in 1959. This increase helped to expand the housing market, created an urgent need for services within urban areas and generated a tremendous increase in consumer spending.

As a carry-over of the Depression and war period, the government continued to play a major role in both managing the economy and facilitating urban development. It was during the early part of this period that bureaucratic power peaked and officials such as Robert Moses increased their ability to shape the urban landscape, with the help of federal funding. There was also a general increase in the roles of government at all levels. Thus, US federal expenditure comprised 10 per cent of the GNP in 1940, but 16 per cent in 1950, and 20 per cent by 1970 (at which time state and local government expenditures comprised an additional 11 per cent of GNP) (Yeates 1998).

The urban population in large metropolitan areas also grew rapidly. Between 1920 and 1950 the number of Americans living in cities rose from 50 per cent to 64 per cent, and by 1970 to 73.5 per cent. Most of this growth, however, occurred on the outskirts or periphery of cities. From 1940 to 1950, metropolitan areas grew by 22 per cent, largely in their suburban regions, while the country grew by 14.5 per cent. Southern and western cities were growing through annexation, and a large black migration from southern to northern cities was also taking place. In the meantime, the northern cities were restructured as wealthier white families moved to the suburbs and poorer black and later Latino families were concentrated in the city core. This process was helped by the government policy of "red-lining", among a series of other polices, which essentially kept city regions segregated (Jackson 1985).

The result of these changes created a tremendous need for infrastructure to accommodate new demands in the suburbs and to replace ageing infrastructure in the core. These needs were at cross-purposes politically as the suburban areas incorporated and fought any city growth. As the tax bases of cities shrank with the movement of upper and middle-income families into the suburbs, the urban cores in many cities began to deteriorate. Early policies to promote local level growth included downtown revitalization and slum removal. Both sets of policies aided by federal dollars exacerbated rather than relieved the problems.

Decisions about improving the water supply were made within the context of rapid urban growth and increasing water consumption per capita. At the same time, the number of waterborne-disease outbreaks decreased rapidly. From 1920 to 1956 there were an average of 25 outbreaks per year, but between 1950 and 1956 no deaths were reported, and water supply management focused on increasing supply. New appliances such as automatic dishwashers, washing machines and air conditioners, in the expanding middle-class population, increased consumption. For example, the automatic dishwasher increased per capita consumption of water by as much as 38 gallons per day. Critics began to portray water supply challenges as part of the "Effluent Society" (Melosi 2000).

As large cities typically already had built their systems, much of the national expansion took place in smaller cities and towns. In 1945, there were approximately 15,400 waterworks in the US supplying about 12 billion gallons per day to 94 million people. By 1965, there were more than 20,000 waterworks supplying 20 billion gallons per day to approximately 160 million people.

With old concerns "solved", new challenges, which only emerged during the previous period, took on crisis proportions. For instance, with the decrease in the number of waterborne diseases, water quality attention after 1945 focused increasingly on chemical contaminants from industries. These environmental and ecological issues were increasingly popularized by books such as *Silent Spring* (1962) by Rachel Carson, *The Quiet Crisis* (1963) by Morris Udall, *The Sand County Almanac* (1966) by Aldo Leopold, *So Human an Animal* (1968) by Rene Dubos, and *The Closing Circle* (1971) by Barry Commoner. The general idea within these texts was that the human race was failing as stewards of the planet and, in particular, for the appropriate care of ecosystems. Many suggested radical change was needed in political, social and economic structures in order to sustain life on Earth.

Biological and chemical assessments of water purity, the nature of waste and the various effects of pollution and harmful human activities replaced the traditional emphasis on public health. Constructed upon a holistic theory, revolving around ecological thought, the new perspective broadened the attack on pollution. Thus, the dominant understanding of environmental problems shifted. As Hays (1987) suggests, after World War II a fundamental change in public values in the US stressed the quality of the environment. Thus, by the 1970s, the dominant environmental themes related to aesthetic and ethical values, which had overtones of nineteenth-century natural theology, and combined with the normative implications of modern ecological science (Caulfied 1989). The new consciousness focused on the relationship between the physical environment and living (including human) organisms, and was called the "new ecology" (Melosi 2000: 281).

One example of the change in attitude toward chemicals in the environment was demonstrated in concerns over air pollution. With years of industrial by-products accumulating in the urban air, a new set of crises emerged. While air pollution had been an issue since the turn of the century,[20] it was brought to the forefront of civic consciousness by the effects of the disasters of, first, Donora, Pennsylvania in 1948, where almost half of the town's 14,000 residents became ill and 20 died, and then by the London fog in 1952, where the death toll climbed to over 3,000 (American Lung Association 2001).

Besides air pollution, a number of other environmental crises appeared. These included the oil spill off Santa Barbara, California, Cleveland's Cuyahoga River fire,[21] the death of Lake Erie and the NYC Hudson River fish kills[22] among numerous others. Together these issues combined and set in motion greater political and ecological awareness that culminated with the inauguration of "Earth Day" (1970).

In response to the new environmental concerns, federal and state governments came to play greatly increased roles in financing and regulating sanitation systems and industrial discharges into the air and waterways. In 1955, the federal government began to regulate environmental quality through the Air Pollution Control Act. This Act only promoted voluntary responses from states, and was only a beginning. The role of the federal government in regulating environmental quality would increase with time. Greater regulatory control and initiatives were taken in many locales to increase the recycling of solid wastes and to manage disposal in a more environmentally benign fashion. During the late 1960s and early 1970s, a set of federal laws were passed that became the seven pillars of the early part of the period (Mazmanian and Kraft 1999; Portney 1990b),[23] which formed part of a command and control regime that dominated that period. They were supplemented by a number of environmental protection and conservation laws focused on land development.[24]

The Golden Age was followed by the Landslide Era, in which global growth was interrupted by recessions and the global slide into instability and crises (Hobsbawn 1996). The last quarter of the twentieth century is associated with a marked decrease in the share of international dominance held by US corporations, a rise in the economic power of Japan, Southeast Asia and the Europe Union, the establishment of a highly integrated world financial market facilitating rapid international capital flows, considerable transnational economic growth–recession instability, the emergence of new products and process technologies, the rise of the service economy in OECD nations, and fluctuating but ever increasing flows of trade and investments among nations. Politically, the fall of the Soviet regime allowed the US to lay claim to be the sole superpower. These changes led to the emergence of a more integrated world economy in which US economic leadership is no longer as strong as it was in the immediate post-World War II decades. The rate of the country's population growth has continued to decline and the ageing population has become significant as elsewhere. Recent population growth in the US is driven largely by immigration.

Core cities continued to depopulate and many cities in the Northeast and Midwest of the country saw their economic bases erode. Metropolises developed multiple centres, including self-contained communities on the periphery, while non-metropolitan growth challenged traditional suburban expansion. Nationwide, southern and south-western cities grew at the expense of older urban centres. The globalizing economy translated into lost industrial employment and increased competition amongst cities for investment and tourist trade. Cities responded by, first, undergoing painful restructuring, from which some have not yet recovered. Those that did recover focused their economic activity on servicing the growing global economy. These post-industrial centres have become the homes of transnational corporation headquarters. As industrial development was no longer the determinant of urban growth, the second response was that cities increasingly found alternative ways in which to promote themselves to investors, This included the building of expensive recreational facilities and conventions centres, providing tax and other incentives and rebuilding their downtown areas to make them more appealing to tourists. Many of the services once provided by governments were privatized. These developments help to entice service industry workers who were looking for cities with a better quality of life and high levels of entertainment. The result of the new types of development has been called "splintered urbanism", as a complex array of various infrastructure, including shopping malls, skywalk cities, secure housing developments, international hotels at traffic interchanges, make the landscape difficult to comprehend if not manage (Graham and Marvin 2001). Importantly, these privately owned infrastructure networks now bring essential services previously provided by governments.

The "new ecology", as a scientific understanding of environmental problems, came to play an important role in helping to formulate environmental challenges, promote responses and provide backing in political struggles. Because the scope of ecology was so broad, however, it splintered into various subdivisions, and from them a variety of fields have emerged. These include population genetics, conservation ecology, agroecology, systems ecology, ecological economics, environmental science and environmental health, among others.

Ecological issues had pushed science far beyond conventional knowledge on many questions about which the evidence was either limited or mixed (Hays 1987). Risk, in its many forms, became the primary means of evaluating the interface between humans and the biogeophysical world. In terms of popularizing environmental issues and helping to fashion environmental regulations, scientific research provided tools for assessing risk or offering new technologies to control pollution. At the same time, it did not uncover unanticipated environmental hazards, but rather introduced a level of complexity not previously realized. Increasingly, understanding environmental science, and the potential threats it revealed, required sophisticated training.

Through these studies a new set of environmental challenges, immune from previous attempts to control them, emerged. This set included burdens that affected entire regions, or the globe itself. Because of the surge in industrial, transport and other activities, greenhouse gas emissions have risen steadily. These emissions are believed to have increased the global surface temperature of the Earth by 1 degree Fahrenheit during the past century, with accelerated warming during the past two decades. In 1997, an international meeting of the Parties to the United Nations Framework Convention on Climate Change developed a protocol to reduce greenhouse gas emissions in Kyoto, Japan. But, although the US is the world's largest producer of greenhouse gas emissions, it has not signed onto the agreement. Recently, President Bush offered an alterna-

tive plan to reduce emissions by an estimated 4.5 per cent over 10 years –
a level dramatically lower than the estimated 33 per cent mandatory reduction sought by the Kyoto agreement.

The national decision to ignore the issue, however, has been under-
mined by a growing number of cities within the US that have taken it
upon themselves to attempt to lower their emissions. For example, the
Mayor of Seattle has promoted the US Mayors' Climate Protection
Agreement that invites cities from across the country to take actions to
significantly reduce global warming pollution. As of June 2005, 165
mayors from 37 states representing 35 million citizens have signed onto
the agreement. Moreover, of the 500 local governments worldwide involved in the International Council for Local Environmental Initiatives
(ICLEI) Climate Protection Campaign, 150 (30 per cent) are located in
the US; cities within the country are acting on their own to address global
warming, demonstrating political concern for the issue at the local level
not seen at the national level.

Despite these actions, however, global warming and other ecosystem-
threatening challenges related to high energy use and emissions are far
from resolved. Further a number of challenges remain dominant including: non-point-source pollution; consumption-related waste, emissions
generation and quality of life issues, such as green space, noise; new
chemicals and toxics; persistent and bio-accumulative toxics (for examples of these issues in a current environmental agenda of a major city,
see New York Conservation Education Fund 2001). These are the types
of environmental challenges that will be among the major risk factors the
country has to address during the coming decades.

Summary

The evolution of urban environmental challenges in the US over the
course of the last 150 years demonstrates several important patterns:
(1) Environmental problems shifted from those related to supplying
    water for health and fire fighting to sanitation (removing wastewater)
    and waste management, then to water and air pollution, and finally to
    complex chemical pollutants and ecologically threatening issues im-
    pacting on the regional and, in some instances, the global community.
(2) Often the "solutions" for one set of problems created an unintended
    set of other problems that emerged at a later date.
(3) As cities developed, they increasingly impacted on larger geographi-
    cal spaces: first, the search for adequate water supplies often forced
    cities to look beyond their borders; second, as wastewater was
    pumped into rivers, those downstream were negatively impacted;
    third, as industrialization proceeded, trans-boundary air pollution in-

creased in intensity and scale; and fourth, as consumption increased, the areas required to supply cities with their daily needs increased – thus cities have expanding ecological footprints.

(4) The shifts in environmental challenges occurred in a serial manner with minimal overlap allowing for a "first things first" approach to problem solving.

(5) The factors influencing the shifts in environmental burdens include economic growth, technological change, theoretical understanding of issues, and the institutions and politics of the environment.

Overview of the telescoping of the urban environmental transition with a focus on the Asia Pacific region

The argument of this chapter is that urban environmental experiences of Pacific Asia cities are significantly different from those encountered in Western cities and that this difference relates, in large part, to the telescoping of environmental transitions in the Asia–Pacific region. There are a number of reasons why the telescoping of environmental transitions has occurred and these relate to the changes in driving forces for environmental change. High on the list of important factors to consider are the prevailing conditions in the region from the 1960s to the turn of the century. These include: (1) the position of countries and cities at the time of their independence from imperial powers was that of large, rural, poor countries without advanced technologies; (2) their rapid development strategies that embraced globalization and the availability of newer technologies; (3) the centralized political systems that encouraged globalization and development often at the cost of social welfare programmes and environmental issues, but that kept a positive political climate for investments; (4) large and, at the time, untapped natural resources that could provide the basic exports and material needs for rapid growth in the early stages; and (5) large populations that placed emphasis on education and advancement through urban lifestyles. Together these factors promoted rapid growth and the time–space telescoping of various transitions, including those related to the urban environment.

In order to demonstrate this trajectory, the third section of the chapter presents two levels of analyses. The first is at the national level. Here, a cross-sectional study of 1995 environmental conditions in nations is presented by income (GDP per capita). Three proxies – namely, access to water, sulphur dioxide (SO_2) emissions per capita and carbon dioxide (CO_2) emissions per capita – are used to simulate the rise and fall of the different sets of environmental challenges. When measures are standardized and superimposed, the resultant chart looks very similar to that posed by Smith (1993) in terms of the health risk transition.[25] These data are then further used to produce the percentage of urban population currently experiencing various environmental risks. Asia Pacific nations are then mapped onto the graph to demonstrate their sets of environmental concerns. The second level of analysis is based upon a similar type of study, using the UNCHS city database (1998). Here again, a set of measures is used to demonstrate relationships between environment and increasing income (in this case City Product per Capita). This analysis shows that at the city level some of the same relationships found at the national level hold (although the analysis is limited to providing proxies only for the first two sets of agendas).

The distribution of environmental challenges in nations throughout the world

An analysis of data for three different types of environmental burdens supports the main points of this argument and presents an environmental development scenario.[26] As the three functions in Figure 1.2 demonstrate, there is a relationship among the following variables, all calculated for 1995: GDP per capita with percentage non-access to safe water, SO_2 emissions per capita and CO_2 emissions per capita. These data, taken from World Resources Institute and the World Bank, demonstrate the outlines of the urban environmental transition. Each function represents

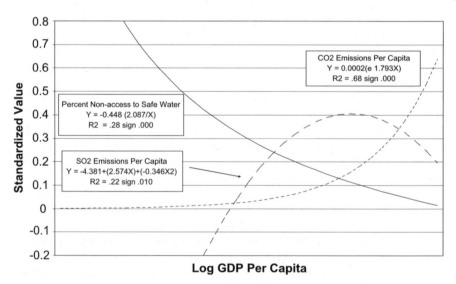

Figure 1.2 The Environmental Transition Data 1995 (*Source*: Marcotullio, Rothenberg and Nakahara, 2003)

a significant relationship between wealth (in this case income) and the three sets of environmental issues.

The figure demonstrates that the problems experienced by low-income cities, such as access to safe water, have an inverse relationship to wealth. These issues make up an important part of the "brown agenda" and include the most critical environmental problems facing Third World cities. They include more than just lack of safe water, but also inadequate waste management and pollution control, accidents linked to congestion and crowding, occupation and degradation of sensitive lands, and the interrelationships between these problems (Bartone et al. 1994). The one variable, access to safe water, is used here as a proxy for the entire group of variables. The trend of increasing access with wealth arguably approximates the trajectory for all the other variables within this agenda, although there are dissimilarities between variables and they do not match each other exactly in terms of appearance and dissolution.

In general, these types of problems decline with increasing affluence only to be replaced by those associated with rapid industrialization, such as SO_2 emissions. This group of environmental challenges makes up the "grey agenda", which includes environmental challenges associated with industrialization and motorization. Some of the specific measures might include certain air pollutants such as SO_2 total suspended particles (TSP) and water pollutants (as measured, for example, by chemical oxygen demand (COD)), and levels of phosphorus, among others. In terms of these largely chemical inputs, Robert Kates (2000) suggested that in the US between 1970 and 1996 the levels of some air and water pollutants decreased (with the exception of nitrogen oxides, which remained constant). He noted the drop in particulate matter and lead from the air and phosphorus from water (see also Table 1.2). The curve for SO_2 emissions per capita by income demonstrates the "inverted U-shape" of the environmental Kuznets curve. This relationship represents the decreasing environmental quality associated with rapid development, followed by increasing environmental quality, once some turning point is reached. This turning point is believed to be a function of increased environmental regulations, as also experienced in Japan (Sawa 1997) since these types of pollutants, largely from the end-of-pipes, declined in intensity with increased controls.

The next set of burdens includes non-point source, consumption related (such as CO_2 emissions and waste production) and persistent chemicals, among others. Increased CO_2 emissions within cities are the partial result of an increase in automobile ownership. Waste production is a result of urban lifestyles and increased consumption on almost all levels. The trend in the relationship between these variables and wealth increases exponentially, although, some relationships may have seen an

Table 1.2 Percentage change in US air quality and emissions, 1980–1999

Pollutant	Air Quality 1980–1999	Emissions 1980–1999
Carbon Monoxide (CO)	−57	−22
Lead (Pb)	−94	−95
Nitrogen Dioxide (NO$_2$)	−25	NA
Nitrogen Oxide (NO$_x$)	NA	1
Ozone (O3 1-hour standard)	−20	NA
Ozone (O3 8-hour standard)	−12	NA
Volatile Organic Compounds (VOC)	NA	−33
Particulate Matter (PM10)	NA	−55
Sulphur Dioxide (SO$_2$)	−50	−28

Sources: American Lung Association (2001); EPA, latest findings on National Air Quality: 1999 Status and Trends.
Notes:
NA = not applicable.
Long-term trends for PM2.5 are not yet available.
Air quality estimates are actual measurements of pollutant concentrations in the ambient air at monitoring sites across the country.
Emissions estimates are based on actual monitored readings and engineering calculations of the amounts and types of pollutants emitted by vehicles, factories and other sources.

s-curve type development (for a decline in CO_2 emission intensity with income, see Hayami 2000).

Each set of relationships were then standardized and superimposed on a single chart to demonstrate an estimation of their relationships to each other. When these three curves are overlaid one upon another they resemble the relationships defined by the urban environmental transition. From these estimations it is further possible to approximate the share of the urban population experiencing different types of environmental risks. That is, the points where the curves meet mark shifts in the types of environmental challenges and, when related to GDP per capita, a very rough estimate, based upon national urban population levels produced by the UN, of the number of people living under the conditions can be calculated (Table 1.3).

These educated guesses are important in that they demonstrate that the majority of the world's urban population (over 50 per cent) is living under conditions of at least two sets of burdens and that over 20 per cent are living under conditions of all three types of burdens. This may be why it has been difficult to separate out the relationship between affluence and environmental conditions in many cities. If these estimates represent any measure of reality there are a number of different types of problems experienced within cities simultaneously. These figures also demonstrate

Table 1.3 Estimated urban population living under various environmental conditions, 1995

1995 GDP Category (US$)	Environmental Challenge	Total Urban Population (thousands) (N)	Share of Total (%)
<467.74	Lack of water and sanitation ("brown" issues)	456,985	17.8
>467.75 and <1,071.52	Rising industrial pollution ("grey" issues), and significant "brown" issues	518,812	20.3
>1,071.53 and <3,981.07	High "grey" issues, rising modern risks ("green" issues) and "brown" issues	526,315	20.6.
>3,981.08 and <14,125.3	High but decreasing "grey" issues, rising "green" issues	296,993	11.6
>14,125.3	Largely "green" issues	613,480	24.0
		147,610	5.8
Total global urban population		2,560,195	

Source: Peter J. Marcotullio, Sarah Rothenberg and Miri Nakahara (2003).

that less than a quarter of the world's urban population is living under conditions largely related to the "green" agenda. This agenda, however, is increasingly the basis of the sustainable development mandate from which came, for example, the compact city model. Lastly, the figures also show that a significant percentage of the world's urban population (18 per cent) is living in conditions dominated largely by the "brown" agenda. This share may seem small, but the reader is reminded that these are not the only ones living under this type of situation. Rather, the curves demonstrate that 38 per cent of the world's urban population is living with significant "brown" issues, together with another 20 per cent living in cities that have yet to resolve them completely. The point, however, is that of this total 58 per cent, 40 per cent are also experiencing multiple burdens.

Finally, when Asian countries are mapped onto the curves, a picture of their current environment set of burdens emerges (Table 1.4). Most of those nations are found in the areas associated with multiple burden levels. These have been called "environmental risk overlaps" and translate into a set of mixed risks (Smith 1990; Smith and Lee 1993). That is, in an urban setting, when there are significant amounts of both traditional and modern risks occurring at the same time, several new classes of problems emerge. For example, Smith (1993) points out that about 40

Table 1.4 Selected Asian countries categorized by predominant set of environmental challenges (1995)

Environment Issue	Country Name	Income Level (US$)	Urban Population (thousands)	Non-access to Clean Water (per cent)	Per Capita CO_2 (tonnes/capita)
Predominantly Brown Issues	Bhutan	171.90	111	37	0.00
	Nepal	197.20	2,193	18	0.00
	Vietnam	275.80	14,362	51	0.02
	Cambodia	276.40	1,413	93	0.00
	Lao PDR	360.60	988	59	0.00
Significant Brown Issues, Rising Gray Issues	China	571.10	362,394	4	0.12
	Indonesia	1,003.10	70,301	31	0.07
High Gray Issues, Decreasing Brown Issues and Rising Green Issues	Philippines	1,093.50	36,940	12	0.04
	Thailand	2,868.30	11,693	9	0.13
High, but Decreasing Gray Issues, Rising Green Issues	Malaysia	4,235.90	10,806	6	0.23
	Republic of Korea	10,142.20	35,168	0	0.36
Predominately Green Issues	Hong Kong	23,463.70	3,321	0	0.22
	Singapore	25,156.20		0	0.84
	Japan	40,846.10	97,950	0	0.39

Sources: WRI Database (2000–2001); UN Urban Population Prospects (1999).

per cent of the people in China still use crop residues for cooking, creating a traditional risk in homes (smoke from burning bio-fuels). Now, however, the pesticide residues left on the crops compound the problems associated with the traditional forms of cooking and heating. These risk overlaps are important for large populations in rapidly growing developing countries.

Further, in the past, citizens of developing cities might experience environmental risks mostly associated around the home or within the neighbourhood. Currently, as suggested by Table 1.3, however, significant populations are affected by these risks and in addition have to cope with consumption related risks that affect larger areas, such as automobile and river pollution.

Hence this analysis suggests that, at the national cross-sectional level, the urban environmental transition exists, but that it is currently being telescoped such that some nations, particularly those in the middle-income range, are experiencing multiple environmental agendas concurrently.

Global city data and the environmental transition

UN Habitat has collected data and indicators on the state of cities around the world through a network of national focal points and partners.[27] One of the goals of this exercise is to provide a monitoring tool for progress in cities towards implementation of the Habitat II agenda. While a recent study suggested that these data are limited in terms of their usability, it also stated that the effort provides a "significant start" in the march towards their intended goal (May et al. 2000). In a recent effort to provide more reliable and usable data, UNCHS has performed another round of data collection. These new data were used to help define the relationship between increasing wealth (in this case, city product per capita) and a number of environmental measures. A group of measurements is used to represent proxies for the brown (access to water, percentage of waste openly dumped and households connected to sewers) and the "grey agenda" (water consumption and percentage of workforce that uses motor vehicles to get to work), as there were no measures for the "green" agenda.

The correlations of the measures with income are shown in Table 1.5. As expected there is a positive significant association between city production and water consumption, households connected to sewers, and percentage of workforce that uses motor vehicles to get to work. At the same time there is a negative relationship between city product and households without access to water within 200 metres, percentage of waste openly dumped and an impact parameter, child mortality.

Table 1.5 Correlations between city product per capita and environmental conditions

City	Per Capita City Product
Households without access to H_2O within 200 metres	
Pearson correlation	−0.343
N	85
Percentage of waste openly dumped	
Pearson correlation	−0.458
N	76
Child mortality (per 1000)	
Pearson correlation	−0.433
N	84
Water consumption (litres per person per day)	
Pearson correlation	0.407
N	79
Percentage of households with sewer connections	
Pearson correlation	0.444
N	65
Percentage of workforce using motor vehicles to get to work	
Pearson correlation	0.250
N	64

Source: UNCHS Database (1998).

The percentages were graphed to present a picture of how the different associations relate to one another (Figure 1.3). The best fit for these various bivariate functions was linear rather than non-linear. They suggest there is overlap among the brown and grey agendas.

Another way to separate these data is to categorize cities by the level of income and calculate means for each group. Table 1.6 demonstrates that for "low", "middle" and "high" groups the various measures representing different aspects of each agenda change with income as predicted.[28] These trends, however, hide the compressed transition. Another way of demonstrating the overlapping environmental challenges, for example, is to look at individual cities within the various income categories and, at the same time, examine their specific measures for each indicator.

Table 1.7 presents the data for cities in Asia sourced from the UNCHS database. It is interesting to note the fairly high levels of vehicular use in low- and middle-level cities, and the high percentage of wastewater that goes untreated in middle-level cities. There is need for caution in interpreting the numbers as they represent the relationships for all cities,[29]

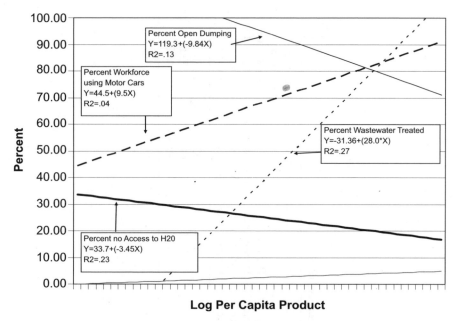

Figure 1.3 Relationship between environmental issues and city product (*Source*: UN–Habitat database 1998)

but they, at least, begin to demonstrate the overlaps that are occurring within the current rapidly developing urban experience in the region.[30] Despite the seeming level of detail in these studies, however, only the most superficial understanding can be garnered without in-depth histori-cal analyses. Unfortunately, size restrictions do not permit this analysis. (For more details on conditions within sets of cities in the region, see Marcotullio 2003, 2005).

Summary

This brief review of environmental conditions within the region, demon-strates the possibility that major urban centres have sets of environ-mental issues related to both their wealth levels and their linkages to the regional city system. Globalization, technological advances and the insti-tutional conditions within rapidly developing Asian countries has brought rapid growth, but has also helped to telescope urban environmental tran-sitions. However, we must obviously approach these analyses with some caution as further empirical evidence to refute or refine the theory is needed. There is much more work that will need to be completed before

Table 1.6 Selected brown and grey agenda measures by city product category, 1998

City Product Category	City Product Per Capita ($)	Brown Agenda				Grey Agenda	
		Per cent Waste Dumped	Households Without Access to H_2O w/in 200 m	Per cent Households with H_2O Connections	Per cent Households with Sewer Connections	Proportion of Wastewater Treated	Per cent Workforce using Motor Vehicles to Work
Low (23)	461	56.7	15.5	84.5	47.4	33.3	30.7
Middle (42)	2,728	53.2	5.4	94.6	75.2	64.5	33.0
High (27)	21,162	9.7	0.3	99.7	97.3	90.0	48.2

Source: UNCHS Database (2001).

Table 1.7 Selected brown and grey agenda measures for selected Asian cities, 1998

City Product Category	City	City Product Per Capita ($)	Brown Agenda					Grey Agenda
			Per cent Waste Dumped	Households Without Access to H_2O w/in 200 m	Per cent Households with H_2O Connections	Per cent Households with Sewer Connections	Proportion of Wastewater Treated	Per cent Workforce using Motor Vehicles to Work
Low Income	Ulaanbaatar	505	90.0	9.7	90.0	60.0	98.0	10.0
	Phnom Penh	699	74.0	14.6	85.4	74.6	0.0	87.3
Middle Income	Surabaya	1,225	0.0	5.7	94.3	55.8	0.0	80.0
	Jakarta	1,932	0.0	8.6	91.4	64.8	15.7	–
	Penang	4,237	80.0	0.1	99.9	–	20.0	42.0
	Bangkok	9,553	0.0	–	–	100.0	–	58.7
High Income	Singapore	22,955	0.0	0.0	100.0	100.0	100.0	25.1
	Tokyo	32,350	0.1	0.0	100.0	100.0	–	–

Source: UNCHS Database (1998).

41

a picture representing the complex realities of Asian urban environments comes into focus.

Conclusions

Cities around the world are experiencing a variety of environmental challenges. At the same time, however, those challenges in the developing world are increasingly seen as different from those in the West and unresponsive to Western policies. While cities in the developing world are undergoing an urban environmental transition (in certain respects similar to what the developed world experienced), there are significant differences between the two regional processes. The urban environmental transition of the West accompanied long wave-shifts in economic and technological change, that of the developing world, particularly cities in Asia, are experiencing serious "risk overlaps" as development occurs rapidly over one long wave. The telescoping of the urban environmental transition can be observed in Asia–Pacific cities that are linked to the regional and global economy. There, development has been linked to the types of institutions found in the region, and these very institutions play a role in facilitating the overlapping of environmental burdens.

The context studied in which the telescoping has occurred, however, is limited to nations and cities in the Asia Pacific and certainly may not be uniformly experienced throughout the developing world. Given the uneven way in which globalization has proceeded, a range of modifications to the urban environmental transition can be hypothesized. Cities in Latin America have experienced an "extended transition", while cities in sub-Saharan Africa have experienced something different from both the extended and telescoped varieties. The point is that not all developing world cities are undergoing similar processes and the compressed environmental transition might be one form specific to the Asia–Pacific region. In each case for developing world cities the transition has certainly differed from that experienced by cities in the West.

At the same time, the telescoped urban environmental transition has significant implications for governance of cities in the region. While in the West, environmental problems were tackled in a serial manner, using a "first things first" modality, this luxury is not available for cities in the developing world. Indeed, increasingly these cities are called upon to solve several problems simultaneously. This calls for more integrated environmental policies to tackle cross-scale issues.

Further, the contexts within which environmental problems were solved in the West have shifted through the years. The advantages of these shifts were that different institutional contexts facilitated different types of responses. These advantages are not only unavailable to current rapidly developing governments, but the governments of the developing world are encouraged to take on policies related to the current challenges of the developed (i.e. privatization, volunteerism, etc.). Whether these policies will work in the developing world context remains an open question.

The urgency of finding solutions to these issues is backed by the certainty that the burdens of the developing world are larger than any nation is able to handle on its own. While integration of policies is often called for, there is currently no clear understanding of how this should be accomplished. In view of the lack of understanding of how to integrate policy responses, more research and practical experience is needed in this area. In this regard, several interesting and positive efforts are currently underway, including the Millennium Ecosystem Assessment, a global assessment of the world's ecosystems, that was launched by UN Secretary-General Kofi Annan in June 2001. This effort promises to provide valuable insight into the trends, conditions, future scenarios and possible response options for urban ecosystems, and is an undertaking that deserves more attention from urban environmental scholars.

Acknowledgements

A version of this chapter was presented at a UNU/IAS Urban Programme workshop entitled "Scaling the Urban Environmental Challenge: From Local to Global and Back", held from 10–12 March 2002 at UNESCO, Paris. During that time, the author received valuable comments and criticisms from all the group's members including (alphabetically) Xuemei Bai, Jeb Brugman, Kristof Bostoen, Priscilla Connolly, Graham Haughton, Amitabh Kundu, Yok-shiu Lee, Gordon McGranahan, Awais Piracha, Joe Ravetz and Jacob Songsore. Andre Sorenson not only provided comments but also copy-edited the document. Masahisa Fujita and those at the Kyoto University Institute of Economic Seminar provided insights that were also incorporated. At UNU/IAS Grant Boyle helped with valuable comments and important edits. The author also received helpful comments from an anonymous reviewer. All errors and inaccuracies remain the responsibility of the author.

Notes

1. US urban history is well documented and provides extreme examples of serial transitions in urban environmental conditions.
2. Variables were grouped into three elements including (1) "financial" (capital interest,

wages, bank deposits, etc.) (2) "mixed character" (the volume and value of foreign trade, etc.); (3) "purely natural" (sector production and consumption levels).

3. In this in-depth analysis of long-term economic trends one finds no convincing evidence to support the notion of long waves, but it does suggest nevertheless that there have been "significant changes in the momentum of capitalist development. In the 170 years since 1820 one can identify separate phases which have meaningful internal coherence in spite of wide variations in individual country performance within each of them". Maddison suggested that the move from one phase to another was governed by exogenous or accidental events that are not predictable.

4. A good example is the use of the automobile to solve problems associated with the use of horses for transportation and work-related activities in cities (see, for example, Melosi 2000).

5. It is the slow response to infrastructure provision that helps to accentuate the overlapping nature of environmental conditions within cities.

6. The exception was New York City where, as the nation's first capital and port of call, the population grew to 100,000 by 1820.

7. While claiming to supply water, which it did to a limited extent and to only the wealthy, it was essentially a cover for banking operations.

8. Crafts (2000) estimates that during this period, the US growth rates were among the highest in the world.

9. The construction of the Erie Canal (1817–1825) solidified the position of New York as the nation's first port and largest and most active commercial city by tying a large portion of the growing Western markets to the city, and provoked Philadelphia, Baltimore and, to a lesser extent, Boston into constructing their own transportation links to the west.

10. Manhattan Island reached 2.3 million people by 1910 and according to Ken Jackson, noted NYC historian, by that time had obtained residential densities higher than any city in the world to that point, and possibly since then.

11. Overnight the city's population grew from a little over 2 million to 3.4 million and it increased in area from approximately 23 square miles to approximately 322 square miles.

12. The Draft Riots occurred over the course of three days in July 1863 in New York City and caused approximately US$1.5 million in damage. Potential conscripts for the war, particularly Irish immigrants, rioted and terrorized blacks. The number of killed and wounded is unknown. In response to the riots the first housing welfare laws within the USA were passed in New York City.

13. Before, the sanitary idea associated environmental problems directly to disease. In the emerging era of bacteriology, filth was not simply a nuisance, it could harbour disease, but at the same time it lost some of its importance as a health hazard. Smells were no longer considered dangerous, and unsanitary surroundings were no longer considered a menace to health. While filth was still a problem, focus turned to disease control and environmental health, which shifted emphasis to controls on residential densities and housing conditions.

14. In 1890, more than 70 per cent of cities with populations exceeding 30,000 had public systems. In 1897, 41 of the 50 largest cities (or 82 per cent) had public water systems. Since most of the urban population was located in the largest cities, it was not surprising that while only 43 per cent of all American cities had public waterworks in 1890, 66.2 per cent of the total urban population was served by public systems.

15. Philadelphia had constructed a large public works system by 1811, followed by New York City in 1842. In 1845, Boston passed an act enabling the city to construct its own waterworks. By 1860 the country's 16 largest cities had basic-large-scale water supply systems of which only four were privately owned (Chudacoff, 1981).

16. The population soared between the mid-1800s and the early 1920s, in large part due to immigration. In 1840 there were 17 million Americans, but by 1910 there were 106 million. This growth was most striking in the cities of the Northeast and Midwest. New York City grew from 200,000 residents in 1830 to more than 1 million in 1860, reaching almost 7 million in the late 1920s when immigration constraints came into effect (Muller 1993). During one day at the height of the immigrant wave in 1907, approximately 12,000 people queued up on Ellis Island for entry into the US and during that year 1.2 million people were received in New York.

17. In 1870, almost no US water was filtered, but by 1880 approximately 20,000 people in urban areas were receiving this service. By 1900 this number had grown to 1.86 million, reaching 10.8 million in 1910 and over 20 million in 1920, accounting for 37 per cent of the entire urban population (Haines 2001). By World War I, most of the largest cities and several smaller ones invested in filtration plants and treatment facilities, making the infrastructure of the modern water-supply system more intricate with these new additions.

18. While the rates of bacteriological and pathogenic diseases decreased, occupational hazards increased. The most poignant moment, and a watershed for those working for safe and healthy workplaces, was the Triangle Shirtwaist fire of 25 March 1911, in which over 140 women (mostly teenagers) died in a workplace fire.

19. At that time, the industrial wastes regarded as the most serious hazards in relation to water-supplies included phenol. Far more dangerous compounds, however, were emerging: coal and petroleum distilling produced benzene, toluene and naphtha; lead waste from crushing and smelting operations began to find its way into watercourses, creating serious industrial hygiene problems; arsenic, used in paints, wallpapers and pesticides, were becoming widespread; and steel mills produced between 500 million and 800 million gallons of pickle liquors and other acids annually.

20. In cities, the smoke from burning coal dirtied clothes and buildings and was a health threat to citizens. Unlike drinking-water contamination and sewerage, however, many people viewed smoke in positive terms as a sign of progress and industrial productivity. Because smoke abatement could threaten business profits and workers' livelihoods, opposition came from both business and labour (Stradling 1999). As a result, by 1940, no states, and only 52 municipalities and three counties, had air pollution control legislation (Portney 1990a, Table 3–1: 29).

21. More than 200 tons of chemical, petroleum and iron wastes, plus domestic sewage, were discharged daily into the Cuyahoga. In 1959, the river burned for eight days and during the summer of 1969 it caught fire again, engulfing two railroad trestles.

22. Across the country, it was estimated that in 1963, there were 7.8 million fish killed by discharges into US water bodies (2,200 miles of river and 5,600 acres of lakes).

23. The seven federal laws included the Clean Air Act (1970), the Federal Water Pollution Control Act Amendments (1972), the Safe Drinking Water Act (1974), the Resource Conservation and Recovery Act (1976), the Toxic Substances Control Act (1976), the Federal Insecticide, Fungicide, and Rodenticide Act (1972) and the Comprehensive Environmental Response, Compensation and Liability Act (1980).

24. Two important examples are the National Environmental Protection Act (1969) and the Wilderness Act (1964).

25. Kirk Smith has been working on "The Risk Transition" in the area of health for over 10 years. His studies are related to "The Epidemiological Transition" of Adbel Omran, but aimed toward facilitating environment risk assessment and management. He has also studied the process of "risk overlap", during which traditional risks are joined with modern risks (see, for example, Smith 1990; Smith 1995).

26. Satterthwaite (2000: 28) has suggested that statistics such as those used in this study may

be inaccurate, claiming "Many statistics do not measure what they claim to measure. Even where they are accurate and wide ranging, they still fail to tell the whole story: they abstract the problem from the local context". Hardoy and Satterthwaite (1986) have also suggested that statistics used as the basis for commenting on urban change in the Third World are not reliable. These authors are concerned with the possibility that these measures will encourage policy makers toward strategies beyond the influence of local communities. We share this interest and present these data analyses with "a grain of salt". At the same time, however, the results demonstrate the need for a multi-scale approach in tackling urban environmental problems, in part complementing the concerns of those focused on possible data inaccuracies. Further, our concern has also prompted the three-tier analysis rather than relying on only national level or even city level data.

27. Originally, data from 237 cities in 110 countries were presented at the Habitat II Conference in Istanbul, 1996. These data were largely 1993 data. Recently (2001), Habitat has assembled a newer dataset for 232 cities in 113 countries within 6 regions for 1998. These data were used for this set of analyses.

28. See also McGranahan et al. (2001), Table 2.4 (pp. 25) and Table 2.5 (pp. 29) for a similar set of analysis using UNCHS 1993 data.

29. A number of dummy variables ("Asian/non-Asian", "North/South", and "industrialized/non-industrialized" did not provide additions to the r-squares or higher significance levels to the regression tests.

30. These data also demonstrate their limitations. Anyone that has been to Jakarta or Bangkok knows that not all waste is treated and that not all households have water connections.

REFERENCES

American Lung Association (2001) "Trends in Air Quality", Best Practices and Program Services.

Asian Development Bank (1997) *Emerging Asia: Challenges and Changes*, Hong Kong: Asian Development Bank/Oxford University Press.

Bartone, C., Bernstein, J., Leitmann, J. and Eigen, J. (1994) *Toward Environmental Strategies for Cities: Policy Consideration for Urban Environmental Management in Developing Countries*, Washington, DC: World Bank.

Bennett, J.W. and Dahlbert, K.A. (1990) "Institutions, social organization and cultural values", in (eds) B.L. Turner, W.C. Clark, R.W. Kates, J.F. Richards, J.T. Mathews and W.B. Meyer *The Earth as Transformed by Human Action, Global and Regional Changes in the Biosphere over the last 300 Years*, Cambridge: Cambridge University Press, pp. 69–86.

Berry, B.J.L. (1997) "Long waves and geography in the 21st century", *Futures* 29: 301–310.

Bruchey, S. (1992) "The sources of the economic development of the United States", *International Social Science Journal* 134: 531–548.

Burgess, R., Carmona, M. and Kolstee, T. (eds) (1997) *The Challenge of Sustainable Cities: Neoliberalism and Urban Strategies in Developing Countries*, London: Zed Books.

Caulfied, H.P. (1989) "The conservation and environmental movements: an historical analysis", in J. P. Lester (ed.) *Environmental Politics and Policy: Theories and Evidence*, Durham: Duke University Press, pp. 13–56.

Chudacoff, H.P. (1981) *The Evolution of American Urban Society*, Englewood Cliffs: Prentice-Hall.

Crafts, N. (2000) "Globalization and growth in the twentieth century", in *World Economic Outlook, Supporting Studies*, Washington, DC: International Monetary Fund.

Cronon, W. (1991) *Nature's Metropolis: Chicago and the Great West*, New York: W.W. Norton.

Dicken, P. (1998) *Global Shift: Transforming the World Economy*, London: Paul Chapman.

Friedmann, J. (1986) "The world city hypothesis", *Development and Change* 17: 69–83.

Friedmann, J. and Wolff, G. (1982) "World City Formation: an agenda for research and action", *International Journal of Urban and Regional Research* 6: 309–343.

Graham, S. and Marvin, S. (2001) *Splintering Urbanism – Reviving Resistance*, London: Routledge.

Haines, M.R. (2001) "The urban mortality transition in the United States, 1800–1940", in NBER Working Paper Series on Historical Factors in Long Run Growth, Cambridge, p. 20.

Hardoy, J.E. and Satterthwaite, D. (1986) "Urban change in the Third World: are recent trends a useful pointer to the urban future?", *Habitat International* 10: 33–52.

Hayami, Y. (2000) "From confrontation to cooperation on the conservation of global environment", *Asian Economic Journal* 14: 109–121.

Hays, S.P. (1987) *Beauty, Health and Permanence: Environmental Politics in the United States, 1955–1985*, Cambridge: Cambridge University Press.

Held, D., McGrew, A., Goldblatt, D. and Perraton, J. (1999) *Global Transformations, Politics, Economics and Culture*, Cambridge: Polity Press.

Hobsbawn, E. (1996) *The Age of Extremes: A History of the World, 1914–1991*, New York: Vintage Books.

Jackson, K.R. (1985) *Crabgrass Frontier: The Suburbanization of the United States*, Oxford: Oxford University Press.

Johnston, R.J., Taylor, P.J. and Watts, M.J. (1995) *Geographies of Global Change: Remapping the World in the Late Twentieth Century*, Oxford: Blackwell.

Kates, R.W. (2000) "Has the environment improved?", Editorial in *Environment* 42(5).

Knox, P. and Agnew, J. (1998) *The Geography of the World Economy*, London: Arnold.

Kondratieff, N.D. (1979) "The long waves of economic life", *Review* II: 519–562.

Lo, F.-c. and Marcotullio, P.J. (eds) (2001) *Globalization and the Sustainability of Cities in the Asia Pacific Region*, Tokyo: UNUP.

Lo, F.-c. and Yeung, Y.-m. (eds) (1998) *Globalization and the World of Large Cities*, Tokyo: UNUP.

Maddison, A. (1991) *Dynamic Forces in Capitalist Development: A Long-Run Comparative View*, Oxford: Oxford University Press.

Marcotullio, P.J. (2003) "Globalization, urban form and environmental conditions in Asia Pacific cities", *Urban Studies* 40: 219–248.

—— (2005) "Time-space telescoping and urban environmental transitions in the Asia Pacific", Yokohama: United Nations University Institute of Advanced Studies.

Marcotullio, Peter J. and Lee, Yok-Shiu (2003) "Urban environmental transitions and urban transportation systems: a comparison of the North American and Asian experiences" *International Development and Planning Review* 25(4): 325–354.

Marcotullio, Peter J., Rothenberg, Sarah and Miri Nakahara (2003) "Globalization and urban environmental transitions: comparison of New York's and Tokyo's experiences", *Annals of Regional Science* 37: 369–390.

May, R., Rex, K., Bellini, L., Sadullah, S., Nishi, E., James, F. and Mathangani, A. (2000) "UN Habitat Indicators Database: evaluation as a source of the status of urban development problems and programs", *Cities* 17: 237–244.

Mazmanian D.A. and Kraft, M.E. (1999) "The three epochs of the environmental movement", in D.A. Mazmanian and M.E. Kraft (eds) *Toward Sustainable Communities, Transition and Transformations in Environmental Policy*, Cambridge: MIT Press, pp. 3–41.

McGranahan, G., Jacobi, P., Songsore, J., Surjadi, C. and Kjellen, M. (2001) *The Citizens at Risk, From Urban Sanitation to Sustainable Cities*, London: Earthscan.

McShane, C. (1999) "The revolution in street pavements, 1880–1924", in G.K. Roberts (ed.) *The American Cities and Technology Reader: Wilderness to Wired City*, London: Routledge/Open University, pp. 107–116.

Melosi, M. (1999) "Refuse pollution and municipal reform: the waste problem in America, 1880–1917", in G.K. Roberts (ed.) *The American Cities and Technology Reader*, London: Routledge and Open University, pp. 163–172.

—— (2000) *The Sanitary City: Urban Infrastructure in America from Colonial Times to the Present*, Baltimore: Johns Hopkins Press.

Miller, B. (2000) *Fat of the Land: Garbage of New York, The Last Two Hundred Years*, New York: Four Walls Eight Windows.

Monkkonen, E. (1988) *America Becomes Urban: The Development of US Cities and Towns 1780–1980*, Berkeley Ca: University of California Press.

Muller, T. (1993) *Immigrants and the American City*, New York: New York University Press.

New York Conservation Education Fund (2001) *New York City Citizen's Guide to Government and the Urban Environment*, New York: NY Conservation Fund/NY League of Conservation Voters.

Portney, P.R. (1990a) "Air pollution policy", in P.R. Portney (ed.) *Public Policies for Environmental Protection*, Washington DC: Resources for the Future, pp. 27–96.

—— (ed.) (1990b) *Public Policies for Environmental Protection*, Washington DC: Resources for the Future.

Riis, J.A. (1972) *How the Other Half Lives: Studies Among the Tenements of New York*, Williamstown: Corner House.

Sassen, S. (1994) *Cities in a World Economy*, Thousand Oaks CA: Pine Forge Press.

Satterthwaite, D. (2000) "Official figures do not tell the whole story", *Habitat Debate* 6: 28–29.

Sawa, T. (1997) *Japan's Experience in the Battle against Air Pollution*, Tokyo: Committee on Japan's Experience in the Battle against Air Pollution.

Schaeffer, R.K. (1997) *Understanding Globalization: The Social Consequences of Political, Economic and Environmental Change*, Lanham: Rowman and Littlefield.

Smith, K. (1990) The risk transition. International Environmental Affairs 2: 227–251.

——— (1993) "The Most Important Chart in the World", in UN University Lectures, UNU Public Forum Series, 4 June 1993, Tokyo: UNUP, p. 28.

Smith, K.R. (1995) "Environmental hazards during economic development: the risk transition and overlap", in E.G. Reichard and G.A. Zapponi (eds) *Assessing and Managing Health Risks from Drinking Water Contamination: Approaches and Applications*, Fountain Valley CA: National Water Research Institute, pp. 3–14.

Smith, K.R. and Lee, Y.-s. F. (1993) "Urbanization and the environmental risk transition", in J.D. Kasarda and A.M. Parnell (eds) *Third World Cities: Problems, Policies, and Prospects*, Newbury Park, Sage Publications, pp. 161–179.

Stradling, D. (1999) *Smokestacks and Progressives*, Baltimore: Johns Hopkins University Press.

Tarr, J. (1999) "Decisions about wastewater technology, 1850–1932", in G.R. Roberts (ed.) *The American Cities and Technology Reader*, London: Routledge/Open University, pp. 154–162.

Tarr, J.A. (1993) "Urban pollution: many long years ago", in K.T. Jackson (ed.) *The Great Metropolis, Poverty and Progress in New York City*, New York: American Heritage Custom Publishing, pp. 163–170.

——— (1996) *The Search for the Ultimate Sink: Urban Pollution in Historical Perspective*, Akron OH: University of Akron Press.

Yeates, M. (1998) *The North American City*, New York: Longman.

2

Formulating sustainable systems

Yoshitsugu Aoki

1. Implications of the term "sustainability"

The term "sustainability", which emerged only a few years ago, has come to be employed in various situations in relation to urban problems and social affairs. The term originates from the "sustainable development" proposal made in the 1987 report of the World Commission on the Environment and Development (WCED). Sustainable development largely represents the idea that our choices should not be dictated by our present circumstances but guided by the principle of sustainability, so that the interests of future generations are not be compromised. This is a very convincing argument if we consider the future of the natural environment and humanity.

Today, sustainability is integrated with a wide range of issues not necessarily related to the environment, as a concept of maintaining decent conditions in the future. Accordingly, this chapter discusses the concept of sustainability in a broad sense, without confining it to environmental issues. However, there is no clear-cut definition concerning the concept of sustainability itself.

In this context, this chapter will first seek to clarify the concept of sustainability by identifying some of the term's implications. The discussion will then focus on the meaning of sustainability as related to urban land use.

2. Constancy and sustainability

Stone Model

Any clarification of the concept of sustainability requires the identification of the image that we would like to give to that concept. In this connection, it would be useful to highlight the common features of things that are deemed sustainable in our daily life. Giving examples of familiar things will help us understand the discussion intuitively. The nature of a specific object is also easy to define mathematically, although this book does not cover such mathematical definitions in detail.

Anything may be considered as sustainable as long as it never changes. Supposing then that sustainability means unchangeable, in order to measure the relevance of this hypothesis, pick up a stone from the roadside and put it on your desk. Over a period of time no change will be observed – the position and form of the stone will be sustained indefinitely.

We might therefore consider this completely unchanging state as a necessary and sufficient condition of sustainability and call this concept a Stone Model.

However, it is quite inappropriate to conclude that the stone is sustainable just because it does not appear to change and so we can conclude that sustainability must mean something different to the mere "absence of change". It is apparent therefore that although stones do not by nature change, they cannot be considered sustainable. In other words, sustainability refers to something that changes by nature but maintains constancy in certain aspects.

Plastic Bucket Model

The discussion of the Stone Model leads us to think that sustainability implies the existence of certain constant aspects in changeable things. Let us reconsider this by playing in the bathroom as we did in our childhood.

First, using a nail, pierce a hole in the bottom of a plastic bucket (Figure 2.1).
1. Fill the bucket with water and observe the change in water level. It will decline gradually until the bucket becomes empty.
2. Next, turn on the tap to the maximum and pour water into the bucket. This time, the water level will continue to rise as the inflow of water exceeds the outflow from the hole at the bottom. Eventually, the bucket will be filled up and water will spill over the top.
3. Finally, turn off the tap little by little. At one point, the water level will stabilize as the outflow equalizes the inflow from the tap.

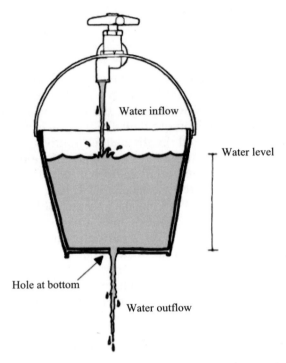

Figure 2.1 The Plastic Bucket Model (*Source*: created by the author)

Technically, the outflow of water from the hole changes with the water level. For convenience sake, however, this example considers the outflow as constant, unless the size of the hole is changed.

The balanced dynamic movement of water inflow and outflow brings about the apparent stability of the water level. We can think of sustainability as referring to this kind of stability.

At this point, let us stop using our intuition and identify the nature of this phenomenon and the relationships that it contains.

In the Stone Model, no factors work to cause change in the stone. In the case of the Plastic Bucket Model, however, inflow from the tap works to raise the water level, while outflow from the hole at the bottom works to lower it. In short, water level is influenced by both a positive factor and a negative factor. The stability of the water level appears only when the two factors are balanced – that is, when the inflow equals the outflow. This is also a general condition that keeps a state constant in a system that experiences dynamic changes. Typically, a variable is constant when the following relational expression holds. This seemingly natural relation is, in fact, very important.

Increment in the variable to be held constant
= decrement in the variable to be held constant

The condition that this relation holds can be called "equilibrium condition". It is relatively easy to express this equilibrium condition strictly in mathematical terms. The state satisfying the condition is called an "equilibrium state".

This concept of equilibrium actually covers the Stone Model: the stone may be considered to satisfy the equilibrium condition as the forces to cause a rightward movement and a leftward movement are both zero. This chapter, however, only considers cases where the above-mentioned equilibrium expression holds when neither the left-side increment nor the right-side decrement equals zero. This kind of equilibrium is sometimes called "dynamic equilibrium" to distinguish it from the cases explained by the Stone Model. Thus, we are brought back to the initial statement of this sub-section: that sustainability implies the existence of certain constant aspects in changeable things.

Urban sustainability as a form of dynamic equilibrium

The discussion so far has brought us to the idea that sustainability means dynamic equilibrium. However, how effective is the concept obtained in the bathroom in understanding real urban issues? This subsection seeks to confirm the effectiveness of the Plastic Bucket Model by associating it with some practical issues.

Sustainability of natural environments

We start our discussion with the origin of the concept of sustainability, that is, natural environmental issues.

How to measure the quality of the natural environment is a very demanding research task in terms of air quality and forest area, for example. For the purpose of our discussion, suppose that the quality of the natural environment is measurable, and that the qualitative level corresponds to the water level in the Bucket Model. What, then, is the equivalent of the outflow from the bottom that brings the water level down? Actually, various factors contribute to the degradation of the natural environment. For example, increased activities in urban areas will result in rising carbon dioxide emissions by increasing the number of automobiles, or in the reduction of forest area through increased residential land development. Bearing those processes in mind, urban activities should be generally considered to lower the quality of the natural environment. On the other hand, what serves to improve the natural environment? In some cases, people contribute to the conservation of global environment by planting trees on wastelands. However, the self-purification capacity of Nature itself is noted more often: plants grow with the blessing of the sun. Thus, the growth and recovery potential of Nature itself should be considered the equivalent of the inflow of water from the tap.

The growth and recovery of Nature is extremely slow. Residential land development will change a forest area in a few years into an agglomeration of concrete buildings with paved streets, whereas it takes decades for Nature to restore the original forest. It should therefore be noted that the growth and recovery potential of Nature only represents a fine stream of water from the tap. Since equilibrium is only possible when the inflow equals the outflow, urban activities have to be maintained at a low level that corresponds to the growth and recovery potential of Nature. A slightly larger hole at the bottom would result in a fast decline in water level. In other words, excessive urban activities lead directly to the degradation of the natural environment.

Actual problems have a more complicated structure. In the Plastic Bucket Model, inflow from the tap stays the same even if the hole at the bottom becomes larger. In the case of the natural environment, however, increased urban activities not only lower the quality of the environment but also reduce the growth and recovery potential of Nature itself, thus accelerating the process of environmental degradation. This process will be discussed later in Section 4.

Sustainability of urban activity level

The previous segment showed that the natural environment would be sustained as long as urban activities stayed below a certain threshold. How, then, may urban activities be sustained at a certain level? This segment verifies the validity of the Plastic Bucket Model from the viewpoint of urban development or decay.

In this case, water level corresponds to the level of urban activities. Inflow from the tap would represent the factors that increase urban activities, whereas outflow from the bottom would correspond to the factors that curtail urban activities. Conversely, turning the tap on to the maximum would correspond to the situation in which urban activities increase due to excessive positive factors until they become untenable, as illustrated by the water spilling over the top. Here again, we would like to identify sustainability with a situation in which the water level is held constant, as positive and negative factors cancel out and bring about a balanced level of urban activities.

3. Stability and sustainability

The discussion on the Plastic Bucket Model in Section 2 has made it clear that there is a common characteristic in what we would like to regard as sustainable: a balance between increments and decrements. That is to say, the existence of equilibrium is the first and foremost condition of sustainability. However, some might still not be satisfied with this definition. This section examines additional conditions of sustainability.

Wok model: automatic correction of slight imbalances

Let us move from the bathroom to the kitchen. Put a golf ball in a wok (i.e. a round-bottomed pan). The ball will soon stabilize at the bottom of the pan. If you flip the ball from its stationary position, it will in time resume its stability at the bottom of the pan. Now turn the wok over and place the golf ball on the top of the upturned wok – this is not an easy task but with care it can be done. One little shake, however, will cause the ball to tumble off the wok (Figure 2.2).

In both cases, the stationary state of the ball indicates the existence of a balance – but there is a difference. In the former case, the ball will resume its position after any movement, whereas in the latter case, a slight movement will cause an infinite deviation from the state of equilibrium.

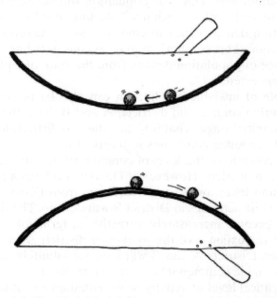

Figure 2.2 The Wok Model (*Source*: created by the author)

We would qualify the former case as "stable" and the latter as "unstable". Thus, equilibrium may be classified as having two differing states – one of stable equilibrium and the other of unstable equilibrium.

A more elaborate description of the above example will provide the conditions to be met in stable equilibrium. Namely, stable equilibrium requires that "any change from a point near equilibrium will lead towards equilibrium". A mathematical indicator is known to determine whether this condition is met.

Seemingly, what we expect from the concept of sustainability approximates to that of stable equilibrium.

Urban sustainability as stable equilibrium

We can also find examples of stable and unstable equilibrium in urban problems.

Take, for example, the population density of an urban residential area. If this area has a relatively small population in comparison to its distance from the city centre, convenience and environmental quality, residents in other areas will be tempted to move into this area. As a result, the density of population will go up gradually. When it exceeds a certain threshold, the attractiveness of this area will begin to decline, partly due to an increase in land prices. In some cases, overcrowding will lead to environmental degradation, inducing some people to move out of the area for better living conditions. Thus, the population will increase until it reaches the density threshold, after which it will decline. Just like the golf ball in a wok, population density tends to converge with a certain level that ensures equilibrium. This equilibrium may be considered as stable, because any divergence of population density from the state of equilibrium is automatically corrected.

An example of unstable equilibrium can also be presented. Suppose two neighbouring commercial districts, A and B, have the same level of activity at the initial stage. That is to say, the two districts have, for example, identical economic conditions in terms of total sales and profit rate. Under these conditions, the level of commercial activity stays almost unchanged in both districts. However, if District A adopts a predatory pricing strategy and takes away some customers from District B, District A will develop while activities in District B will decline. The district thus developed will become increasingly attractive in terms of product choices and prices at the expense of the weakened district. As a result, the disparity between Districts A and B will increase infinitely, leaving District A as the only viable commercial area. It can therefore be understood that the initial identical level of activity represented an unstable equilibrium.

As demonstrated by the above examples of stable and unstable equilibrium, it is the state of stable equilibrium that we expect from sustainability.

4. Closed system and sustainability

In the Plastic Bucket Model, water inflow from the tap does not change regardless of water level. However, the inflow may be altered in response to water level. In the natural environment, it is known that the extent of the growth and recovery of Nature changes according to the quality of the environment. This section addresses the concept of sustainability in such a complex system.

Grazing Pony Model

To understand the complex mechanism of the natural environment, this subsection employs one of the simplest models. Imagine an environment in which ponies are grazing on a vast meadow. Attention should be focused here on the number of ponies (X) and the amount of meadow grass (Y). More grass means more feed for ponies, resulting in a larger number of foals. We assume the number of foals born from a pony per year to be aY. Unfortunately, some ponies will die each year. We assume that ponies die at a ratio of b per head. In this case, the increment in the number of ponies per year may be expressed as follows:

$$\Delta X = aXY - bX$$

Meanwhile, the amount of meadow grass also depends on the number of ponies. As new grass grows from the existing grass, its growth changes in proportion to the current amount of grass. We assume the increment of meadow grass in a year as c per unit. However, grass is eaten by the ponies. Ponies cannot eat when there is no grass, but will eat more when the grass is abundant. Thus, the amount of grass eaten changes in proportion to the amount of grass available and the number of ponies. By assuming a coefficient of d to explain this relationship, the increment in the amount of grass per year may be expressed as follows:

$$\Delta Y = cY - dXY$$

How, then, do the number of ponies and the amount of grass actually change? The two expressions shown above are known as Volterra equations, the mathematical characteristics of which have been studied by many researchers. Let us introduce the fundamental results of technical discussion on this type of equation.

The state of equilibrium has already been defined in dealing with the Plastic Bucket Model. In the present case, however, equilibrium is defined as the state in which no changes are observed in the two variables, that is, the number of ponies (X) and the amount of meadow grass (Y). No change in X means ΔX equals zero in the first equation. The right side must also be zero when the left side is zero. Thus, the equilibrium condition for the number of ponies may be expressed as follows:

$$aXY - bX = 0$$

Likewise, the equilibrium condition for the amount of grass $(\Delta Y = 0)$ may be expressed as follows:

$$cY - dXY = 0$$

The number of ponies (X) and the amount of grass (Y) are both in equilibrium if the above two expressions hold. This condition is satisfied in the following two cases.
(1) $X = 0$, $Y = 0$
(2) $X = c/d$, $Y = b/a$
The first case indicates that neither ponies nor grass exist. Obviously, such a bleak landscape does not fit our image of sustainability. It may therefore be said that only the latter case represents the concept of equilibrium. Namely, it refers to the situation where the number of ponies and the amount of grass both remain unchanged at certain levels.

Expected changes in the number of ponies and the amount of grass from a state of non-equilibrium may be grasped in an intuitive manner.

Draw a plane graph showing the number of ponies (X) on the abscissa and the amount of grass on the ordinate. The latter equilibrium mentioned above may be plotted at the black point in Figure 2.3.

Let us divide the graph into four regions according to whether the number of ponies is above or below the level of c/d and whether the amount of grass is above or below the level of b/a. The four regions may be described as follows:

Region A: the number of ponies exceeds c/d and the amount of grass exceeds b/a,

Region B: the number of ponies exceeds c/d and the amount of grass is less than b/a,

Region C: the number of ponies is less than c/d and the amount of grass exceeds b/a, and

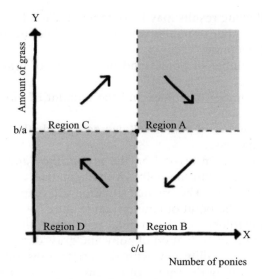

Figure 2.3 Expected trends in the Grazing Pony Model (*Source*: created by the author)

Region D: the number of ponies is less than c/d and the amount of grass is less than b/a.

The increments in the number of ponies (ΔX) and in the amount of grass (ΔY) may be calculated separately for each region. In Region A:

$$\Delta X = aXY - bX = aX(Y - b/a)$$

where a, X and $Y - b/a$ are all positive. Thus,

$$\Delta X > 0$$

Also,

$$\Delta Y = cY - dXY = dY(c/d - X)$$

where d and Y are positive and $c/d - X$ is negative. Thus,

$$\Delta Y < 0$$

The above two results indicate that the number of ponies will increase in Region A while the amount of meadow grass will decrease.

Thus, the following results may be obtained for the four regions:

Region A: the number of ponies will increase, while the amount of grass will decrease,

Region B: the number of ponies and the amount of grass will both decrease,

Region C: the number of ponies and the amount of grass will both increase, and

Region D: the number of ponies will decrease, while the amount of grass will increase.

Showing those trends in arrows on the graph, they appear to move in circles around the equilibrium point. A mathematical elaboration proves this intuition to be right. The original situation will reappear after a circular motion around the point of equilibrium (Figure 2.4).

Hence, any divergence of the original point from the state of equilibrium means that the combination of pony and grass does not come nearer or go further from equilibrium, with the original distance between the two points representing the radius of the circular movement. If the original point is in Region A, it will move into Region B, and then to Regions D and C in that order, eventually returning to Region A. First, the number of ponies will increase and the amount of grass will decrease, but in time, the number of ponies will also decline. Subsequently, however, the amount of grass will begin to rise, followed by an increase in the number of ponies, resuming the initial situation. Thus, the number of ponies and the amount of grass will both experience cyclical fluctuation.

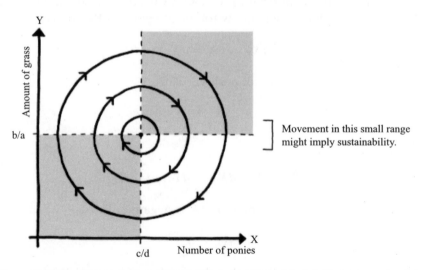

Figure 2.4 Direction of change in the Grazing Pony Model (*Source*: created by the author)

The further the initial point is from equilibrium, the wider becomes the range of this repeated fluctuation. In this context, a small circular movement starting from a point near equilibrium may be regarded as static. Such movement should meet our criteria of sustainability.

This model has another important implication. An increase or decrease in the number of ponies impacts on the increase or decrease in the amount of grass available, which in turn causes an increase or decrease in the number of ponies. The cyclical movement explained above is caused by this interaction between the number of ponies and the amount of grass available. This leads us to conclude that the concept of sustainability includes this type of repetitive, dynamic change caused by a cyclical movement, in addition to the state of stable equilibrium described earlier.

Cyclical structure and sustainability of cities

The Grazing Pony Model implies that minor changes in the number of ponies caused by its interdependence with the amount of grass may meet the criteria of sustainability. This type of phenomenon is often observed in the real world.

Let us start with a brief description of real cities to see how the level of urban activities (X) and the quality of natural environment (Y) are determined. In a city that has not developed to a threshold level, the level of urban activities does not improve easily due to factors such as underdeveloped infrastructure and inconvenience. Beyond the threshold level, however, the efficiency of industrial activities tends to improve with the development of infrastructure. Also, a high-quality natural environment means better river water conditions, which provides advantages in terms of residential land development and plant location. It can therefore be said that the increasing factor of urban activities comprises a high level of urban activities themselves (X) and a high-quality natural environment (Y). However, increased plants and residential land may also be disadvantageous for urban activities. Therefore, the level of urban activities themselves is also considered as decreasing. The increment of urban activities (ΔX) can be expressed as follows:

$$\Delta X = aXY - bX$$

The positive factor for the quality of the natural environment includes first and foremost the growth and recovery potential of Nature. This growth and recovery potential works in proportion to the quality of the natural environment (Y). It is also true, however, that urban development tends to cause the destruction of Nature. Therefore, environmental deterioration is correlated to both the quality of the environment itself (Y) and the level of urban activities (X). The increment in natural environment quality (ΔY) may be expressed as the differential between those positive and negative factors:

$$\Delta Y = cY - dXY$$

The two expressions are identical to those used in the Grazing Pony Model. Therefore, the quality of the natural environment will decline when the level of urban activities exceeds c/d. Also, urban activities will decline when the quality of the natural environment falls below b/a. The level of urban activities may go up above this environmental threshold. Thus, the two elements will repeat their ups and downs.

The validity of the Grazing Pony Model facilitates the discussion of possible measures to promote urban development and natural conservation. Suppose, for example, the level of urban activities and the quality of the environment is plotted at $a0$ in Figure 2.5.

In this case, the natural environment and urban activities will both improve to reach $a1$. From this moment on, however, the natural environment will decline while urban activities will continue to increase. The quality of natural environment approaches to zero. Mathematically

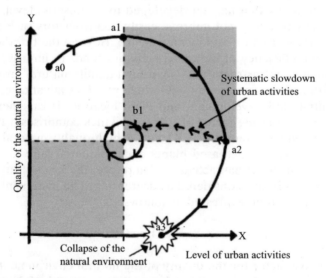

Figure 2.5 Urban activities and the natural environment (*Source*: created by the author)

speaking the orbit of Volterra equation does not cross with the X-axis, but there is the high possibility of becoming zero in the realistically slight change. If the quality of environment becomes zero at $a3$, the level of urban activity will also move toward zero. This means the collapse of the city and the environment. Therefore, measures must be taken to bring the situation somewhere inside the orbit before we reach $a3$. For example, a systematic slowdown of urban activities at $a2$ may avoid the expected collapse by shifting the level to $b1$.

5. Stochastic variation and sustainability

Choice and chance

We have reached the concept of stable equilibrium by using examples as simple as a stone, a plastic bucket and a wok. Some may question how such simple concepts can explain the complicated structure of a city, with a large population and lots of buildings and roads. Indeed, the life of an individual may be in stable equilibrium, but a neighbour may be experiencing a vicious cycle of poverty. Do we have to consider the state of stable equilibrium for each person and element in the city? If so, we may have to deal with an almost infinite number of combinations. For instance, there are only two ways to put a coin on a table: head or tail. However, we can put 10 coins in 1,024 combinations (i.e. the 10th power of 2). This indicates how complicated urban conditions can be. Is stable equilibrium relevant to such complicated conditions? Unlike coins, city dwellers are far from identical.

City dwellers choose their own behaviour. They do not act systematically by simply obeying orders, unless they live in a Hitlerite state.

The multiplicity of elements is one of the fundamental difficulties in considering urban problems. In the Plastic Bucket Model, a simple relationship exists between water level and inflow/outflow. With urban problems, however, it is not clear how individual elements are related to each other. Some problems even occur by chance. For example, a TV report of a restaurant frequented by a celebrity may attract people, leading to the prosperity of the local shopping area as a whole. Thus, the development of a local shopping area may depend on whether one of the customers happens to be a celebrity. Contingencies may be considered likewise simply because they are unpredictable.

Therefore, the concept of sustainability has to assume that individual city dwellers act independently and that some events may occur purely by chance.

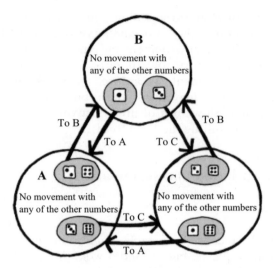

Figure 2.6 The Dice Game Model (*Source*: created by the author)

Dice game model

Imagine a dice game in which people move between three places, A, B and C according to the fall of the dice. The rule of the movement is as follows (Figure 2.6).

At A, two and four indicate a movement to B, and three and six indicate a movement to C. No movement occurs with any of the other numbers.

At B, one indicates a movement to A, and three indicates a movement to C. No movement occurs with any of the other numbers.

At C, two and four indicate a movement to B, and one and six indicate a movement to A. No movement occurs with any of the other numbers.

Participants can freely decide where to start. The number of participants should be as large as possible.

Imagine 30 people start from A, 10 from B and 60 from C and observe the subsequent movements.

Although the random nature of dice prevents us from giving any definite results, a rough estimate can be made using the theory of probability. In this case, the number of people will stabilize at around 25 at A, 50 at B and 25 at C after several trials.

Let us recall the topic of choice and chance discussed in the previous subsection. Although each participant moves back and forth as they choose, the number of people at a given place at any one time does not change significantly. Applying this phenomenon to a city, it may be observed that some aspects of the city do not change as a whole even if individual city dwellers act independently.

Before discussing practical urban problems, we should understand why there is no significant change in the number of people at each place in the Dice Game Model.

The change in the number of people at A, B and C (*Na*, *Nb* and *Nc* respectively) may be calculated as follows.

Inflow to A includes *Nb*/6 persons from B and *Nc*/3 persons from C. Outflow includes *Na*/3 to B and *Na*/3 to C. Therefore, the net increase of people at A is determined as follows:

$$\Delta Na = (Nb/6 + Nc/3) - (Na/3 + Na/3)$$

Likewise, the net increase at B is determined as follows:

$$\Delta Nb = (Na/3 + Nc/3) - (Nb/6 + Nb/6)$$

Also, the net increase at C is obtained as follows:

$$\Delta Nc = (Nb/6 + Na/3) - (Nc/3 + Nc/3)$$

The net increase in each place is zero where *Na* equals 25, *Nb* equals 50 and *Nc* equals 25. This attests to the result presented above that the number of people stabilizes at 25, 50 and 25 respectively. We initially assumed that 100 people participated in the game. With 1,000 participants, the same mechanism will work, resulting in 250 people at A, 500 at B and 250 at C. With only one purely hypothetical participant, we obtain stable percentage shares of the three places. Namely:

$$Na = 0.25, \quad Nb = 0.50, \quad Nc = 0.25$$

Those figures may be considered as the probability that a person finds him or herself in a given place. In this context, the increment of people at A (ΔNa) signifies the net increase in the probability that a person finds him or herself at A. The above expression of the net increase of persons at A actually corresponds to the probability of inflow into A minus the probability of outflow from A. The same thing can be said of B and C. In the final analysis, the condition for no change may be expressed as follows:

Probability of increase at A = probability of decrease at A

Probability of increase at B = probability of decrease at B

Probability of increase at C = probability of decrease at C

This condition may be summarized in one general expression:

Increment of probability of a situation
= decrement of probability of that situation

We can describe this as a "stochastic equilibrium condition", and the situation that satisfies this condition can be called a state of "stochastic equilibrium".

This state of stochastic equilibrium may best approximate the concept of sustainability in a city or society that has numerous aspects and is affected by chance.

Sustainability as stochastic equilibrium

This subsection examines a concrete example of stochastic equilibrium in a city. A city has various kinds of business establishments operating in it in various sectors. These businesses may be newly established, wound-up, or they may even move from one area to another within the city. In this example, we deal with the movement of business establishments within a city. For convenience sake, we have divided a city into three zones: A, B and C. Individual businesses may move from one zone to another for various reasons including convenience. Although it is impossible to predict such movements because each business establishment has its own set of characteristics and circumstances, let us assume that the following statistical law stands.

In five years, a third of the business establishments in Zone A will move to Zone B, and another third to Zone C, with the rest remaining in Zone A.

In the same period, a sixth of the business establishments in Zone B will move to Zone A, and another sixth to Zone C, with the rest remaining in Zone B.

Likewise, a third of the business establishments in Zone C will move to Zone A, and another third to Zone B, with the rest remaining in Zone C.

This is of course an oversimplified model, and the proportion of movements may be described by other more sophisticated models (e.g. the logit model and the spatial interaction model). However, all are based on similar stochastic ideas.

Imagine that 30 business establishments are located in Zone A, 10 in Zone B and 60 in Zone C. Observing the number of business establishments in each zone at five-year intervals, it turns out that the number approaches 25 in A, 50 in B and 25 in C. In this example, Zone B appears to be the industrial centre of the city. Although some businesses will continue to move around, the numbers will become increasingly stable, just as in the Dice Game Model.

In reality, the number of business establishments in each zone stays almost level, barring any substantial economic change or large-scale development projects. This process applies not only to business establishments but also to other urban activities including the number of housing units.

As can be observed in this example, the number of various urban facilities or the level of urban activities in a zone may be stabilized as a whole in a balanced manner, albeit with some internal changes due to various factors. It may be said that this state, namely, stochastic equilibrium, is the closest to what we would expect from sustainability.

6. Sustainability of land use

This section looks at the trend of land use in cities based on the concept of stochastic equilibrium discussed in the previous section.

To simplify, we assume that a city has n lots that can be used for two purposes only: commercial and non-commercial. Then we focus our attention on the i-th lot at a certain point in time. If the lot is surrounded by commercial land, it may be advantageous to transform it into a commercial lot in order to develop an intensive commercial area. Thus, we can determine the relative advantage of using a given lot for commercial or non-commercial purposes according to the use of surrounding lots at a certain point in time. We may call "commercial incentive" the extent to which the use of a lot as commercial land will be advantageous. A given lot has both commercial and non-commercial incentives. If the commercial incentive is greater than the non-commercial incentive at one point in time, then the lot can then be used as commercial land.

As discussed in the previous section, however, city dwellers have free will and each incentive is affected by unpredictable events. In other words, commercial and non-commercial incentives are unevenly distributed.

A rather complicated mathematical discussion on this subject leads to the following results (Aoki 2001 and 2002):
(1) There exists only one state that meets the requirements of stochastic equilibrium.

(2) Under certain conditions, land use approaches the state of stochastic equilibrium.

(3) In that case, a specific indicator exists, the value of which will decrease over time.

The first result – that is, the existence of stochastic equilibrium – seems only natural but is quite important.

The second result is very significant. Although it was not thoroughly discussed in the previous section, stable equilibrium in a stochastic process simultaneously meets the requirements of equilibrium (addressed by the Plastic Bucket Model) and stability (addressed by the Wok Model). With regard to stochastic equilibrium, there could be a concept of stochastic stable equilibrium that indicates a tendency to approach the state of equilibrium. A further sophisticated discussion would make it clear that this may also be regarded as a stochastic version of the Wok Model, and the second result proves that convergence with the state of stochastic stable equilibrium may be observed in some cases.

According to the two results above, a state of stochastic stable equilibrium may be achieved eventually in a hypothetical city that has only two land-use purposes, subject to certain conditions.

Although this example represents an oversimplification of a complicated urban structure, an actual city may approach a stable equilibrium form of land use if certain conditions are met, even when individual city dwellers continue to act freely.

With regard to the third result presented above, it may be proven mathematically that the decreasing indicator equals the reverse value of another indicator, namely, the synergy of land use.

The latter indicator increases when lots used for the same purpose are located nearby and those used for other purposes are located further away, and the indicator declines in the reverse scenario. Looking at the third result in this light, it may be concluded that when certain conditions are satisfied, the synergy of urban land use improves.

This conclusion can be expressed in more concrete terms. Current city planning laws and regulations provide for a specific land use purpose for each zone, thus prohibiting development projects for any other purposes. However, the above-mentioned results imply that the same land use will eventually be concentrated in particular areas under a loose system that allows individuals to select the use of urban lots so that their benefits may be maximized, as long as certain conditions are satisfied. In theory, it would not be an exaggeration to say that zoning may be achieved even without any land use regulations. This naturally developed land use pattern is definitely one of the elements that constitute our vision of a sustainable city.

REFERENCES

Aoki, Yoshitsugu (2001) "The sustainable urban state as a stochastic stable equilibrium and its existence conditions", *Urban Planning Review* 36: 949–954 (text in Japanese).
—— (2003) "The possibility of the zoning without the regulations", *Journal of Architecture and Planning* (Transaction of AIJ) 573: 79–83 (text in Japanese).

3

Sustainable economies and urban sustainability

Makoto Maruyama

The call for the political economy in the broad sense

The economy has two meanings: the substantive and the formal

The word "economy" conjures up for us two different spheres – namely, on the one hand, that of basic human necessities and, on the other, that of money management. The former consists of everyday requirements, such as food, clothes and habitation – in other words, the satisfaction of material needs, including the empirical processes of production, consumption and the disposal of materials. In this sense, this first type of economy is substantive. By contrast, the second type of economy consists of financial control and profit maximization, and is usually defined as the logical process of optimizing scarce resources. In this sense, the second type of economy is formal.

We often merge these two meanings into one when it comes to a market economy. In the market economy, we have to buy goods and services to survive – that is, to sustain our livelihood through the consumption of commodities – and to do this we have to sell our own goods and services. Since this seems so natural to us, we tend to place money-making in the centre of our livelihood. Nonetheless, we also understand the limits of merging these two meanings together. For example, we sometimes earmark money for gift-giving and distinguish it from ordinary commodity consumption. We also earmark some portion of our salary for mortgage and would not spend it for gambling. Gambling might bring us more money, but usually we lose. And when we lose, we may lose our homes and even families.

It is without doubt, therefore, that formal and substantive economies intermingle broadly and complexly in the market economy. Because of the cost factors involved, we usually purchase necessities rather than making them ourselves, and, in order to earn money to make these purchases, we commute to offices and factories. In a modern society, we feel compelled to acquire more money in order to own more commodities. In such a society, the notion of "having" is superior to that of "being" or "doing". Ivan Illich once called the driving force which produces this kind of situation as radical monopoly (Illich 1973). The term "radical monopoly" describes a kind of mode specific to the industrial society where the material wants of human beings must be satisfied by commodity consumption. In such a society, we cannot, for example, freely drink water from a well or brook, but are coerced into buying "coca cola" to satisfy our thirst. Because of this radical monopoly, we have little access to the domain where the notion of "being" and "doing" is recognized as more meaningful than the notion of "having", although we know from our own experiences that radical monopoly deteriorates the quality of life and puts cities into precarious situations.

In this chapter, I will mainly focus my attention on the domestic structures of advanced countries from the viewpoint of rebuilding these countries from within in order to reduce the burden which has been put on the shoulders of the developing countries. I know that there have been ample discussions on international cooperation in terms of economic development and environmental issues. However, these discussions often separate the domestic issues of the developed countries and the international issues in question. My argument here is therefore a kind of complement to these discourses on economic development and environmental cooperation at the international level.

I will first argue that the sustainable economy must be understood not by means of the formal sense of the economy but by means of the substantive sense of the economy. Then I will use the notion of entropy disposal and claim that sustainable economy must be embedded in the ecological environment which has the ability to keep a low-entropy state. Thirdly, I will shed light on the idea of regionalism and illustrate the decentralized social framework which enables the ecologically benign economy. Finally, I will illustrate how cities can be properly placed in a regional society.

The issue raised by Karl Polanyi

According to Karl Polanyi (Polanyi 1957, 1977), an historian and economic anthropologist, people acquire material means by having an impact on the natural environment and/or through relationships of mutual interdependence in order to satisfy their various needs that arise as they engage in their day-to-day lives. That, as a whole, is "economic" in a substantive sense. By contrast, he continues, the process of obtaining the maximum effect by making the best use of a scarce means is "economic" in a formal sense.

Economists, particularly those trained in the neo-classical school, have popularized the formal concept of "economy" by establishing it as the science of "the optimum distribution of scarce resources", and attempting to also explain the substantive concept of "economic" as well through this definition. Polanyi states that it is unjustifiable to generalize in this manner in the first place. The formal "economic" concept is linked to the substantive "economic" concept through the market system. The neo-classical school of economics believes that scarce resources are distributed most efficiently by being bought and sold through market transactions. However, not all material means to satisfy the needs of everyday life are scarce to begin with. Non-market-based economies have devised systematic measures to prevent resources from becoming scarce, such as collective management and production that matches the speed of renewing resources. A specific example is the use of a system called "the commons". This system has been used to manage the land that concerns a local community, and to distribute labour internally within the community through mutual aid.

The market system comes into common use for the first time when land and labour, as the fundamental components of production, are commoditized. However, a problem lies ahead in this course of thinking. Since neither land nor labour are produced as commodities initially, a special process is required to commoditize them.

Historically, the commoditization of land and labour has been established through various turbulent processes. For example, there was the long process known as the "Enclosure Movement", which took place in England over a long period from the mid-fifteenth century to the nineteenth century. A typical example of this was the process by which open farmland and then common land as well were enclosed by landlords, and farmers were driven off these lands and deemed propertyless labourers. In addition, even once the commoditization of land and labour was achieved, they could not be entrusted wholly to free transaction on the market. Polanyi states that if land and labour become objects for exchange on a free market, the fragmentation of residential space and the mobilization of people advances and a state of cultural "vacuum" emerges. Because people could not endure such a state, they placed restrictions on the free trade of land and labour in order to protect themselves. Polanyi calls this the "self-protection of society." And he comes to the conclusion that it is not possible for the substantive "economic" domain to be completely subsumed by the formal "economic" domain.

It is for this reason that Polanyi broadens his subject for examination to the pattern of integrating an economy through a method separate from the market economy. That is, he focuses not on the form of exchange of commodities, but on the market-less society which satisfies people's needs through a form of reciprocity and redistribution. In such a society, the economy does not submit to the maximization of the profits of corporations and the maximization of the use of household income, but is institutionalized in order to fulfil targets that make local people's livelihoods sustainable. Polanyi discovered this pattern of institutionalization in Aristotle's illustration of household management, and thought very highly of it.[1] Polanyi's approach to the human economy eventually called for political economy in the broad sense, that is, the political economy which targets human economies embedded in social nexus and includes market economies as a part of it.

Yoshiro Tamanoi's advocacy of the political economy in the broad sense

Yoshiro Tamanoi, a Japanese heterodox economist who was deeply influenced by Karl Polanyi, was an advocate of political economy in the broad sense (Tamanoi 1978). He began his argument by directing his attention to one of the negative aspects brought on by post-war Japan's high economic growth, that is, pollution and environmental destruction. In the process of high growth, industrial productivity rose dramatically and production increased very rapidly. Accompanying this, industrial waste products containing toxic substances were diffused widely in the ground and the atmosphere and flowed into areas such as underground water sources, rivers and seas. Since these pollutants exceeded nature's ability to purify itself, environmental deterioration advanced cumulatively, causing pollution and environmental destruction. Tamanoi called these negative phenomena of economic growth as the phase of the "self-negation" of industrialization.

Tamanoi attributes the "self-negation" of industrialization to the very nature of the modern industry. According to Tamanoi, the manufacturing industry prior to the Industrial Revolution was still embedded in the agricultural industry, which was part of ecological activities. In other words, humans participated in the ecological chain in order to sustain and enrich it from within. Tamanoi holds that, after the Industrial Revolution, the manufacturing industry was divorced from agriculture and therefore separated from the ecological circuit. The manufacturing industry adopted the use of fossil fuel and mineral resources rather than biomass and thereby acquired the means of producing artificial materials which were not necessarily benign to the natural world.

Unlike the utopian economists, Tamanoi argues for the possibility of re-embedding industry back into the ecological chain by re-establishing a strong relationship between manufacturing and agriculture on the basis of the soft energy path. This actually means the replacement of the oil-dependent civilization with a biomass-based one. The focal point here is not to eliminate the modern industry absolutely but to shift it towards an ecologically benign direction. The originality of Tamanoi lies in his argument that the ecologically benign economy described above requires the notion of Polanyi's substantive economy. Modern economics only deals with market phenomena and/or interprets all economic transactions as market ones. By contrast, Tamanoi sheds light on the substantive meaning of the economy which includes various types of non-market transactions. In fact, he attempts to advocate a political economy in a broad sense, that is, a political economy which sees markets as a part of the broader domain of substantive economy. The main subject of political economy in the broad sense should therefore no longer be merely the market system or the market mechanism but the substantive economy as a whole benignly embedded in the ecological environment. This is what Tamanoi meant by political economy in a broad sense.

Entropy problems and a living system

The "disposal" of entropy

In order to develop further the paradigm of political economy in a broad sense, the notions of entropy disposal and living system must be introduced at the core of its theoretical structure. According to Tamanoi, a living system is "a system that maintains a stationary state through 'disposing of excess entropy', which results from being alive" (Tamanoi 1982: 91). This definition of a living system comes from Ervin Schrödinger's argument on life (Schrödinger 1967: 75–80). By using the notion of the living system, it becomes appropriate to explain ecology as an eco-system in which living systems form a circuit of giving and taking low entropy energy and matter. By definition, each living system produces excess entropy which is diffused in the outer system, namely, the ecosystem. Living systems therefore depend on the ecosystem which absorbs the excess entropy created by each living system and disposes of it further in the outer system, that is, in space.

Entropy is a technical term of thermodynamics and indicates the degree of diffusion of heat and matter.[2] In this chapter, it is used as a rough measurement representing the degree of increase in waste heat and waste matter that is generated when work is executed by consuming energy.

Theoretically speaking, waste heat is heat in an equilibrium state where no movement of heat takes place. In other words, heat cannot be converted to work any more. It is heat that, from experience, we know has no use – in other words, it is useless energy. Generally, a disequilibrium state of heat is required in order to produce work from energy. When there is heat at a state of high temperature and heat at a state of low temperature in a system, the former is diffused and mixed with the latter, and, as a result, heat energy is converted to mechanical energy. Once heat reaches an equilibrium state in the system, work cannot be produced any more. This state is usually called "heat death". In respect of waste matter, it is a mixture of matter that contains microscopic quantities of various substances combined in a disorderly manner. It is material that we know from experience has no use value. In order to perform work continuously, measures must be taken so that waste heat and waste matter do not increase excessively within the system in which the work is executed.

According to the second law of thermodynamics, in an isolated system where there is no input or output of heat and matter, entropy increases steadily and reaches its maximum level, but does not decrease thereafter, when heat and matter are completely diffused. Consequently, if a living system were an isolated system, the system would become filled internally with waste heat and waste matter as a result of activities to stay alive and it would eventually arrive at a state of "heat death". In order to continue living, a system must be open and its waste heat and waste matter disposed of. In the case of the disposal of waste heat in the human body, this occurs when, on the one hand, the body is cooled through perspiring, while on the other hand, it burns energy within and produces heat, thereby producing a disequilibrium state of heat. In the case of waste matter, this is disposed of by discarding it outside the body as excrement. It is in this manner that an open system is able to take in and discard heat and matter, and a living system is able to rid itself of any excess entropy that has been produced. For the purpose of maintaining life, therefore, the discarding of waste heat and waste matter produced within the system must be regarded as fundamental.

What is important here in respect of the planet is that the ecosystem is the intermediary between its living systems and extraterrestrial space in respect of the disposal of excess entropy generated by any living systems. Earth is an open system with regard to heat, but constitutes a closed system with regard to matter. Therefore, although waste heat can be discarded from the global environment by directing it to outer space, waste matter must be disposed of on the planet itself. Micro-organisms decompose organic waste matter and convert it into inorganic matter, which plants then absorb to produce low entropy matter. Animals consume this matter and waste matter is created once again. The ecosystem, which is founded upon this food chain, is as a whole a repeating cycle in which, as it consumes low entropy heat, it converts high entropy matter into low entropy matter and then consumes that. The high entropy waste heat generated by this overall process is discarded into outer space through the circulation of water and air on the surface of the earth.

The focal point now is the ability of the ecosystem to extract high entropy from matter and return it as heat in order to guarantee that the planet maintains a low-entropy state in terms of matter. If the speed of the diffusion of waste matter exceeds the capacity of the ecosystem with regard to entropy disposal, then the entropy problem emerges. Global warming is an example of such a problem.

Entropy problems and the ecological footprint

In the 1990s, the ecological footprint was developed in Canada as a new environmental index. The ecological footprint visually represents the impact of economic activities upon the ecosystem according to land area. Dr. William Rees of the University of British Columbia led the development of this index (Wackernagel and Reeds 1996). According to Dr. Yoshihiko Wada, who participated in its development under Dr. Rees, the ecological footprint is "the total area of an ecosystem, or water and land, needed to perpetually produce resources that a certain region requires and moreover needed to continuously absorb and dispose of waste matter discharged by the region (it is irrelevant whether these exist within the region or outside it)" (Wada 2001: 30). Specifically, this would be the combined area of forested land that supplies forest products, cultivated land (including ranches) that supplies food, land covered by manmade objects, such as cities and roads, forested land that absorbs carbon dioxide produced through energy consumption, and water areas that supply marine products.

The total area of the ecosystem that actually exists within the region is called the region's environmental capacity. By comparing this with the ecological footprint, it is possible to obtain the specific environmental balance of the region concerned. If the environmental balance is in the red, it either means that the region is weakening its own environment or that it is burdening another region's environment with its load.

The Worldwide Fund for Nature (WWF; formerly known as the World Wildlife Fund) calculates the world's environmental balance based on the ecological footprint method. It publicizes the results every year in the form of the "Living Planet Report" (WWF 2002). Table 3.1 uses data on the US, Canada, Australia, Japan, China, India, and the world as whole, taken from the 2002 edition of this report.

Table 3.1 Environmental balance by country (1999)

	EF (total volume) (1 bil. ha)	Environ-mental Capacity (total volume) (1 bil. ha)	Environ-mental Balance (total volume) (1 bil. ha)	EF (per capita) (ha)	Environ-mental Capacity (per capita) (ha)	Environ-mental Balance (per capita) (ha)
U.S.	2.72	1.48	☐1.24	9.70	5.27	☐4.43
Canada	0.27	0.43	0.16	8.84	14.24	5.40
Australia	0.14	0.27	0.13	7.58	14.61	7.03
Japan	0.60	0.09	☐0.51	4.77	0.71	☐4.06
China	1.96	1.32	☐0.65	1.54	1.04	☐0.51
India	0.76	0.68	☐0.09	0.77	0.68	☐0.09
The World	13.65	11.36	☐2.29	2.28	1.90	☐0.38

Notes:
EF is the abbreviation for ecological footprint.
☐ indicates negative values.
In addition, the environmental capacity per capita for the world corresponds to the world's fair allocation of area. It should be noted that since the data by country were obtained by multiplying the per capita figures by the population size, there are slight differences with the total volume figures.
Source: Prepared based on WWF (2002: 21–27).

According to Table 3.1, in 1999 the land area (including water surface area) that was able to support human economic activity for the world as a whole totalled 11.36 billions hectares, while the ecological footprint totalled 13.65 billions hectares. This means that the environmental balance for the world as whole was in the red by 2.29 billions hectares and that humans were burdening the planet with a load that exceeded the environmental capacity by approximately 20 per cent. In respect of the world's environmental capacity per capita – which corresponds to the world's fair allocation of area per capita – the figure in 1999 was 1.90 hectares, whereas the world's ecological footprint per capita was 2.28 hectares. According to the "2002 Living Planet Report", the world's environmental balance moved out of the black and into the red in 1979. The fact that the environmental balance of the world as a whole is in the red means that the planet's ecosystem is burdened with a load that exceeds its ability to renew and indicates that the deterioration of the ecosystem is advancing.

Table 3.1 gives us several very interesting facts. For example, of the regions with comparatively large ecological footprints per capita, the environmental balances for the US and Japan are in the red, while the same for Canada and Australia are in the black. It is clear that the US and Japan would not be able to maintain their current standards of living without burdening other regions with large environmental loads. Meanwhile, there are cases where the environmental balance is in the red even for regions with comparatively small ecological footprints per capita, such as in the case of China. Since the environmental capacity per capita differs greatly, depending on the region, there are countries such as Canada and Australia that can maintain both high living standards and healthy ecosystems. By contrast, China, with a relatively low standard of living, burdens other regions with large environmental loads and, at the same time, is also weakening her own ecosystems. Of course, if calculations are based on a per capita basis for the population, China's environmental balance deficit would be about one-eighth the size of Japan's deficit. In the case of India, its ecological footprint per capita was as small as 0.77 hectares in the 2002 report, whereas its environmental balance deficit per capita was almost nil. In other words, the ecological balance of India appears to be maintained well below the level of the world's fair allocation of area per capita.

Regarding the method of reducing the load on the environment, we must first think about the reduction of the world's ecological footprint as a whole in relation to the level of the world's ecological capacity. The ideal solution would be to make the ecological footprint per capita in each nation closer to the world's fair allocation of area per capita. For example, Japan, whose ecological footprint per capita is 4.77 hectares, must reduce this by 60 per cent.[3] The United States, whose ecological footprint per capita is 9.70 hectares, must, in the same way, reduce this by 80 per cent. These nations are urged to put into effect immediately the industrial policy which is suggested in the principle of "Factor Four" or "Factor Ten"[4] (Sachs et al. 1998). By contrast, China and India, whose ecological footprints per capita are 1.54 and 0.77 hectares respectively, can still increase their ecological footprints without creating an additional burden on the Earth.[5]

Let us delve into the implication of the ecological footprint approach further. By definition, the regional ecological footprint itself does not mean the burden to the regional ecosystem in question. For example, most of the ecological footprint of Japan consists of the ecological load generated in other nations who export their products to Japan. In the process of producing these products, the exporting nations contaminate their own environment. The ecological footprint approach therefore tries to make clear who is actually responsible for the environmental degradation that takes place across the world by attributing the relevant parts of ecological footprints to the product-importing nations. What the ecological footprint approach does not necessarily clarify is the environmental contamination which would take place in a nation when imported products become garbage in the nation. Also the ecological footprint approach does not illustrate clearly the ecological degradation created in the manufacturing and exporting regions. Despite these limitations, the ecological footprint approach does provide us with a rough idea of where the nations of the world are situated in the complex nexus of wastes.

The organic and the mechanical modes of production

Humans must be able to identify their position within the ecosystem in order to maintain the quality of their own lives. A lifestyle based on hunting and gathering in a primitive society could be identified as part of the food chain as it reproduced its stock, namely, the ecosystem. Traditional slash-and-burn agriculture was also an activity that acquired agricultural products as it repeatedly recycled a secondary natural environment. In addition, even in the most advanced forms of agriculture there exist many systems that embed the production activities of humans within the ecosystem in a manner that conforms with the ecosystem, such as the paddy field cultivation practised in Japan.

For now, let us term a mode of production embedded in the ecosystem an *organic* mode of production, and, in contrast, let us term a mode of production that produces goods independently from the ecosystem a *mechanical* mode of production.[6]

The greatest difference between the organic mode of production and the mechanical mode of production is the method of disposing of high entropy matter that is generated as a result of production. Regarding waste heat, both systems of production are able to dispose of it by making use of the circulation of water and air. Needless to say, this circulation of water and air is mediated in various ways by the ecosystem. However, as far as waste matter is concerned, to date the mechanical mode of production has not yet possessed a system for disposing of this waste on its own. Although the production of low entropy matter as industrial products results in by-products that are high entropy matter, the mechanical mode of production in the twentieth century lacked a process for treating these by-products and converting them back into low entropy matter. Since it is not easy for the ecosystem to dissolve these artificial products (such as plastics) automatically, high entropy matter is almost irreversibly accumulated in the environment.

Industrialized societies of the twentieth century have mechanized not only industry, but even agriculture as well, as symbolized by the use of farm tractors together with chemical fertilizers and pesticides. As we saw in the previous section, twentieth-century style industrialization, that is, industrialization based on a mechanical mode of production, has already created excess wastes beyond the Earth's capacity for disposal. In the twenty-first century, we must therefore restore the balance between the mechanical mode of production and the organic mode of production so that we can maintain the sustainability of the ecosystem which is a prerequisite for all living systems.

Rebuilding the industrialized society from within

The regionalist influence on the political economy

When mentioning Aristotle's self-sufficient economy at the start of the chapter, the idea was introduced that a local community should in principle be self-sufficient and that commerce should be allowed only to the extent of supplementing self-sufficiency. Modern economies are the very opposite of this. First of all, they adhere to the principle of market economy and are structured so that regional self-supply is permitted only to the extent of supplementing the market economy. In the past, the economy of a local community was based on what that community could produce and provide. The wealth of the community was mainly distributed either through mutual compensation between members of the community or by redistribution via the community's central body, and limits were placed on the role fulfilled by the market. However, modern economies are mainly based on commercial production that is premised on free market transactions and it is regarded desirable to have a globally open market.

The point that one should note here is that nowadays the market place is solely a centre for commercial exchange and a system for such transactions; it is not designed or organized as a place for living in the way that a local community was (and, in some regions, still is). When a local community attempts to acquire affluence by utilizing the modern market system, it should first focus on the self-reliance of the community – that is, how the community will protect and maintain its living space – and adhere to the principle of self-sufficiency. Otherwise, it is bound eventually to fall into decline as a depopulated society that has been exploited by the market.

In respect of the Ricardian model of free trade based on the international division of labour in terms of production, the model obviously needs certain qualification. First of all, it remains at the abstract level and does not necessarily fit into the concrete situation. For example, Adam Smith, the author of *The Wealth of Nations* and the founder of the economic theory of free trade, using England as a model, worried about the capitalist development that actually took place in many places in Europe. His concern was mostly the order of investments (Smith 1979: 360, 395–396). According to Smith, the local and regional economies would develop endogenously and retain self-sufficiency when monies were invested, first, in agriculture, secondly, in industry and, thirdly, in commerce. By contrast, he considered that when monies came into villages and towns from outside via merchants, as happened in many places in Europe, regional self-sufficiency would be lost and regional economies would be exploited by international markets. Although Smith was one of the strong promoters of international free trade – a theoretical standpoint reinforced by David Ricardo – he argued that the free-trade system must be supplemented by indigenous local economic development. It is quite interesting to note that Friedrich List – who created the theoretical foundation for protectionist trade to promote German industrialization in order to counter free trade – agreed on this point. Both Smith and List presumed that an increase in surplus agricultural production was a prerequisite for the prosperity of industry and commerce. Furthermore, they believed that the advancement of industry and commerce would, in turn, promote the growth of agriculture.

Both Smith and List understood that economic growth would not in reality be realized exactly according to their ideals, but rather that commerce entering a region from outside would give rise to industrialization and that investment in agriculture was apt to be left to the last. By the 1840s, when List had written *The Principles of National Economy* (List 1841), the essence of the problem of capitalist agriculture, arising from the gap between the ideal order of investment and the practical order of investment, had indeed become clear. List asserted the importance of investment in agriculture and proposed the creation of medium-sized conglomerate farms with livestock-raising as a major sideline in his book *The Farmland System* (List 1842). Society did not directly accept what List advocated, but his basic concept came to be reassessed along with the rebirth of regionalism in Europe, and this will be discussed in the next section. It should therefore be emphasized here that the political economy and economics of European origin have directly or indirectly supported regionalism, and, in return, European regionalism has strengthen the tie between agriculture and industry at the regional level. In other words, the restoration of the balance between the organic mode of production and the mechanical mode of production has gradually become the common sense of European countries.

Revival of regionalism in Europe

It is generally regarded that those European Union (EU) countries which have achieved currency unification, have realized the creation of a huge unified market. There is certainly nothing wrong with that view itself when considered from the point of trades, investments and the movement of labour forces within the region which can be carried out freely. However, at the same time, a generous development programme has been established to promote the economic self-reliance of comparatively less-developed, or undeveloped, regions within the EU, which provides financial assistance to those regions.[7]

The EU's regional development programme is based on planning and implementing individual industrial policies for each assisted region. One example is the Basque region of Spain, which prospered in the past because of its iron and coal industries. While maintaining the conventional systematic course of its economy through its traditional cooperative association method, it is searching for a regional economic system that can accommodate a globally open economy. For instance, it is moving in the direction of permitting the introduction of market-oriented corporations and the associations' surplus funds are being allocated for investment outside the region.

The same is true even in Britain, which is not participating in the current currency unification. Wales is a British region, similar to the Basque region, that had achieved industrialization through its iron and coal industries, but then stagnated because it was slow in modernizing and changing direction. However, by cultivating a high quality labour force and providing labour for relatively low wages, Wales is promoting the active introduction of foreign investments, such as from the automotive industry, and the localization of such industries in the region.

Where does the strength to achieve regional economic self-reliance within the EU countries come from? This is probably not unrelated to the tradition of regionalism that has existed in Europe since the Middle Ages. As is widely known, under the reign of feudal lords (*Herrschaft*), various cooperative organizations (*Genossenschaft*) representing the respective interests of people of the various ranks of society and with the power to negotiate with feudal lords existed in western Europe from the twelfth century. These included groups of aristocrats opposing a lord, the knighthood and autonomous towns free of a lord's power. A dual social relationship of control and cooperation was formed in each region. Through this relationship of tension between these two components, regional unity would gradually develop to form a territorial state. Here the term "territorial state" refers to a state whose unit was territory governed by a sovereign. The state would include various local societies, such as provincial towns and rural villages, and these societies would be interwoven with one another to form a regional cluster. The state would thereby be unified into a broad overall region. The previously mentioned examples of the Basque region and Wales are also broad overall regions with towns and rural villages firmly established within them. Despite the experiences of bourgeois revolutions, modern Europe has carried on the tradition of regionalism from the Middle Ages.

Table 3.2 Grain self-sufficiency of major industrialized countries

(Unit: %)

Country	1984–86	2000	Country	1984–86	2000
Australia	297	279	Britain	105	112
France	221	191	Austria	125	103
Canada	146	165	Switzerland	53	62
U.S.	109	133	Netherlands	30	29
Germany	106	126	Japan	22	25

Note: Although Japan is virtually self-sufficient as far as rice is concerned, its overall grain self-sufficiency is low because it relies on imports to satisfy the majority of its demand for other grains.
Source: Based on "Global Economic Diagram", 1995/96 edition, pp. 229–232; and "Global Economic Diagram", 2002/03 edition, pp. 213–215.

When considered in this way, one can easily understand why grain self-sufficiency is rising in all modern European countries. According to Table 3.2, even a small country such as Switzerland has a rate of grain self-sufficiency of about 60 per cent. Neighbouring Austria is over 100 per cent self-sufficient. Britain also exceeds 100 per cent, and France is 200 per cent self-sufficient. In other words, countries as small as or smaller than Japan have established within themselves a system of internal circulation of grain, the basic staple. It is only upon first establishing this that trade at the EU level, and further up at the global level, supplements this broad regional economic circulation. Furthermore, one should not overlook the fact that this kind of regional circulation system, although temporarily weakened by the process of nineteenth-century industrialization, is once again being strengthened amid economic growth following World War II.

The reason for this is that European regionalism makes much of the balanced development of agriculture and industry as the basis of economic growth at the regional level. This is because an unbalanced system of commerce that promotes the division of labour between agriculture and industry extending beyond regional borders, much less beyond continental borders, does not fit well with the European cultural climate.

Modern Japan as a highly centralized nation

A truly serious problem for Japan's local communities has been that the ability to design a place for living has been taken away from regions in the process of modernization since the Meiji Era. Instead of developing the regional network of cities and villages, industrialization centred on large cities has promoted Japan's centralization of administrative power by leading the country in the direction of firmly establishing a hierarchical structure that subordinates provincial markets to a central market.

In contrast to this, we have the unique history of the development of the market economy during the Edo period (Maruyama 1999). In the early part of the Edo period (in the seventeenth and eighteenth centuries), privileged merchants promoted the formation of a hierarchical market structure that placed the Osaka market at its pinnacle. However, in the late Edo period (in the eighteenth and nineteenth centuries), the advancement of the growth of provincial markets resulted in the development of a network of trade among regions. This gave rise to the progress of commerce in the form of interconnections between the pyramid-shaped market structure centred on Osaka and the regional network-type market framework. Of great interest is the fact that, as though to coincide with this, in the late Edo period there was an increase in the number of feudal clans issuing paper money whose currency was limited to within the boundaries of the respective fiefs. Of course, there were not a few cases of clans that had issued paper money as a makeshift measure to ride out the financial difficulties they had fallen into. However, the destiny for those was to vanish in a short period of time because the market refused to accept them. The background as to why clan-issued paper money in general came to be regarded as "bad money" is recorded in "Hansatsu Sodo (Paper Money Uproar)", a record of the problem-fraught clan-issued paper money, and is also burned into the memories of many people. The reason can be pinned on the fact that during the approximately ten-year period from when the Shogunate opened a Japanese port to the outside world to the time of the Meiji Restoration, one after another small and weak clans began issuing paper money as though seeking refuge in it.

As illustrated above, there was the rich legacy which had prepared the rapid advance of modernization in Japan since the Meiji Era. This legacy was the establishment of a vertical and horizontal market structure and the advancement of regional economies during the Edo period. Modern Japan's market economy, however, exhausted the legacy and did not renew it for future generations. As a consequence, modern society in Japan lost its control over the hyper-centralization which was imposed upon regional societies by the national government.

"The age of the provinces" and the "decentralization" in Japan

Since the latter half of the 1970s, the "age of the provinces" has been a popular movement sweeping the times in Japan. Until that time, the anti-pollution movement, the consumer movement, and other such movements, were citizen action campaigns for self-protection against individual and concrete problems that posed a threat to the lives of the citizenry. The "age of the provinces" was first proposed by some prefectural governments which opposed the national development plans of further centralizing the national economy. The campaign of the "age of the provinces", encouraged by local residents, aimed at more general problems, such as the regeneration and invigoration of regional society based on the idea of decentralization. One can say that it is an active campaign by the provincial governments and local people to attempt to recreate everyday life itself by means of reducing the dependency upon outer markets which are connected with global free trade. The "age of the provinces" movement by local residents has become firmly fixed as an effort to "nurture towns and villages" undertaken by municipalities and citizens' groups throughout the nation. And this movement continues to this day. In this respect, the concept of regionalism has become established in Japan too during the past quarter century after a century-long suppression of decentralization.

However, the fact that the scope of this nurturing of towns and villages is mainly limited to the municipal level is not without problems. In Japan, there are regional states, between the nation state and municipalities, called prefectures. Prefectural governments are supposed to combine together the clusters of municipalities within each prefecture as parts of regional unity. In fact, however, the actual prefectural governments are subordinate to national government due to the concentration of budgets to the national government.

At present, as part of the national policy to promote the decentralization of power, the empowerment of municipalities by means of expanding city limits and consolidating towns and villages is under way throughout the country. The national government seems to use prefectural governments mostly as instruments to empower regional societies. However, the main objectives of these moves are to increase the efficiency of municipal governments in order to save budgets and to reduce numbers of public servants as a whole. Therefore, they are not necessarily efforts that will lead to strengthening the network of municipalities in order to promote self-sustaining economic systems at prefectural level. In fact, the national government seems to view each municipality as a unit of the divisions of labour at the global level. In other words, the government seems to force municipalities to compete with each other in the global markets without depending on governmental protection. In a sense, this is a utopian blueprint based on the Ricardian model of the division of labour among regions.

As the European case in respect of the revival of regionalism suggests, there should be some kind of regional economic policy which ensures the independence of local economic development from the force of making regional economies as parts of global division of labour. However, since Japan lacks a history of regionalism, it is not able to rapidly adopt a European-style economic structure. Nonetheless, when taking into account the country's extremely low rate of grain self-sufficiency (see Table 3.2), Japan must acquire decentralized industrial policies by strengthening prefectural polities.

The rise of "zero emissions"

Japan has looked down upon itself as "a country poor in resources", because it has relied excessively on oil and mineral resources while allowing its rich stock of renewable resources, such as forests, cultivated land and water, to go unused. Utilizing these idle resources and reducing the imported products made from oil is essential to lowering the ecological footprint per capita of the Japanese people. Unfortunately, in the process of industrialization Japan did not develop the technology or systems necessary to support production methods that utilized biomass. In addition, after World War II the country adopted a progressively Americanized lifestyle, thus those natural resources along with their ecologically benign lifestyle were lost. In order to revive this lost lifestyle, we must revise our market-oriented society which depends so much on the scarce resources of oil and minerals. This is not to imply that we must go back to a feudal society based on agriculture and embedded in the local ecology. Such an anachronism is impossible. What is suggested here is that the balance between mechanical and organic modes of production should be regained by developing new technology that preserves the ecosystem while, at the same time, substituting oil and mineral products with more eco-friendly biomass products.

The crucial question here is how can Japan rebuild its own society, the society which would institutionalize the development of ecologically benign technology. Towards the end of the twentieth century, the concept of "zero emissions", where waste matter is not produced, appeared. The zero emissions concept attempts to link multiple and different production processes within a circle of waste matter transmission – or recycling. The circle is not limited to processes within one plant or corporation, but can be formed among multiple corporations or production activities. For example, at the Kokubo Industrial Complex in Yamanashi Prefecture, used paper discarded within the complex is recycled into toilet paper to be used within the complex (the first effort in the endeavour). Waste plastic is transformed into solid fuel and supplied to cement factories as fuel. Food scraps from employees' cafeterias are used to make fertilizer, which is then supplied to local farmers. Furthermore, the organic vegetables cultivated on these farms are purchased by the cafeterias. In this way a matter circulation system has been established built around the industrial complex.

However, one point that demands attention is the fact that it is necessary to input low entropy energy in order to make zero emissions work. The process of converting matter in a state of high entropy into a state of low entropy produces high entropy waste heat. If one is to be precise, the process of producing low entropy energy already generates waste heat and produces carbon dioxide as well. The process for producing fuel cells is by no means an exception in this respect. Furthermore, it is near to impossible to completely return waste matter that has become high entropy matter to its former state of low entropy. Even paper and plastic undergo advancing deterioration as they are recycled repeatedly and must finally be disposed of as garbage. In the same way that a perpetual engine is not possible, 100 per cent zero emissions would be impossible to achieve in reality.

Furthermore, there is an additional problem when considering a system that can produce close to zero emissions: if one attempts to complete the waste matter circulation cycle for just the manufacturing sector without linking it to agriculture, the cycle will remain separated from the ecosystem and merely end up forcing the disposal of excess entropy upon the ecosystem in a one-sided manner. If that burden becomes too great, it will destroy the ecosystem. In order to restore the damaged ecosystem and raise its entropy disposal ability, the organic mode of production must be incorporated into the zero emissions cycle, and one must have production activities by humans that simultaneously maintain and enrich the ecosystem. Consequently, one can give high marks to the Kokubo Industrial Complex's endeavour for its integration of an agricultural products cycle into its zero emissions system.

The example described above is an experiment carried out by one particular industrial complex to which surrounding farms are attached, but it illustrates the type of social framework that could accept a zero-emission production circuit at the core of its regional economy.

Sustainable cities

The city as a life-world

The requirements to make substantive economy sustainable have been noted earlier. For one thing, efforts must be made to restore the balance between the organic mode of production and the mechanical mode of production. Another point is that it is necessary to establish independent and organized regional societies as intermediate structures between local communities and the global market economy. In this section, I will illustrate the position that cities will have in this reorganization of regional areas and how cities should be restructured internally.

First of all, it is important to invigorate economic transactions between city dwellers and the areas surrounding cities. For example, in Tokyo the centralized structure must be changed whereby products from the provinces are gathered at a central market in Tokyo and then returned back to the provinces again for distribution. Provincial cities must be given the cohesive power that promotes the internal circulation of products within their respective regions. Furthermore, it is for this reason that the cities themselves must diversify their needs. The diversification of needs here does not necessarily mean the expansion of commercial consumption. These are needs that do not fit into the existing framework of commercial production and consumption. They are, for example, home-made foods and handicrafts that do not appear on the market but are confined to the homes of farming families that are not on the conventional tourist routes. It is important to exhibit the need for these goods and activities which, in the past, city dwellers routinely made and participated in, but have now lost touch with.

Recovering what has been lost might at first appear to be an unrealistic ideal. However, at present city dwellers can connect with producers only through market transactions. For them, therefore, it would involve forming entirely new relationships with producers in their immediately outlying areas. Thus, it is conceivable that this would be the beginning of the circulation and use of regional resources within the region in a way that had not previously existed. In this case, cities would be not merely areas of consumption, but could also bolster their function as bases for providing information and producing high added-value products through a more efficient processing of regional resources.

The city as a life-world must be an area that is close to the point of production and that also contains a network of production sites within.

The network of local communities and the city

A regional society would certainly not be complete as just a network of local communities and a city at its core. It must be open to the country as a whole, and to the world. The problem is what form this "openness" would take.

There are two types of trade between regions – namely, vertical trade, where resources are exported and processed products are imported, and horizontal trade, where one type of product is exchanged for another. If we consider the basic system to be the use of resources and the circulation of goods within a broad region, then horizontal trade should be the basic trade system and as such it is necessary to endeavour to export highly processed products manufactured within a region to outside areas. Only on such efforts is it then desirable to supplement any needs that internal circulation cannot satisfy through vertical trade.

So far, what has been noted above, has completely ignored the effect that investment has in a market. Consequently, it would appear to the eye of an investor, whose primary objective is asset investment, to be a naive argument that is worse than useless and absurd. For instance, let us take the case of the construction of a golf course by developing mountain land close to a village in order to take advantage of the popularity of golf and compare this with preserving the mountain land and building a local industry that utilizes biomass supplied by this mountain land. From the standpoint of an investor, his objective would best be achieved by purchasing the mountain land, developing the golf course and the sale of all the membership rights to the golf course. In this respect, developing a condominium and selling all the units instead of construction of a golf course would amount to the same thing. If the prices of golf course membership rights and condominium units were stable on the market, these kinds of development projects would be attractive to an investor.

By contrast, it would be extremely difficult for an investor to find any meaning in preserving mountain land. First of all, it would involve efforts such as research and development for the product that utilizes the mountain land, the provision of the equipment and expertise to build the local industry and the pioneering of a sales route for the biomass product, all of which would require starting from zero. Furthermore, if the investor needed to obtain the complete cooperation of local residents who had been utilizing the mountain land up to that point, this kind of investment would not have a chance of success in this respect because the risk would be too high.

However, if one reconsiders the effects of conserving mountain land from the standpoint of residents living in the region, one can see other possibilities. The value of the land would not just be limited to it being a conservation area. By considering it broadly as a means of promoting ecotourism as well as a place that supplies renewable resources, the decision to maintain the mountain land in its natural state would be justified by its contribution to the region's economic activity. Research on mountain land that made use of local knowledge could lead to the development of technology involving the use of bio-resources and so promote the building of a local industry. If interest in the region's resources heightened as a result of the diversification of the needs of city dwellers and the demand for biomass products increased, this kind of economic activity could gain impetus and the growth of biomass-related industry could lead to an increase in jobs as well.

A production system of this kind would probably be built around small and medium-sized businesses. Hence, it would not only be less efficient than a big industry, but would probably require a greater amount of labour. Depending on the region, this could result in a labour shortage which, in itself, could be a good thing. Because the problem of regions suffering from depopulation is not one of a shortage of labour, but one of a shortage of jobs. Therefore, one can suggest that building small and medium-sized regional industries could be a means of keeping people in the region.

From the viewpoint of an organic mode of production, the Tokyo area has rich resources of forests. In the Okutama district of Tokyo, there is a surplus stock of mature Japanese cedar trees. Nevertheless, cedar is rarely used in the construction of urban housing. Overseas materials and new building materials account for most of the market and there are few opportunities for the commercialization of Tokyo's cedar resources. This is not simply a matter of the price of overseas materials being cheaper. Rather, it is the result of a lack of investment in the development of the technology required to design and produce cedar housing suitable for Tokyo's climate.

If the thinking were that housing in Tokyo could be built using local cedar resources as the basic material, thus supplementing shortages of wood from other regions, and replacing imported or new building materials, then it might be possible to design a local industry-related structure that valued the regional Okutama cedar resources. However, in order to execute such a plan, the Tokyo Metropolitan government would have to possess the authority to provide the necessary subsidies and special loans to the people who maintain Okutama's forestry businesses and who would build housing using Tokyo cedar. If we suppose that for the time being the existing prefectures – or the framework for prefectural alliances that will probably be studied in the future – are selected to be the prototypes of such self-sufficient regions discussed here, then the national government must transfer all related funds to them thus making these financial resources totally independent from national finances.

At the same time, local people must install the necessary administrative tools to enable such decentralization. One of the most promising tools for reinforcing regional self-sufficiency is local and regional currencies.

Local and regional currencies

Local and regional currencies are already circulating inside local communities and regional societies, mediating the transactions among people who not only trade goods and services through markets but also trade their home-made products and aids in the residential areas.

Over the past few years, Japan has been experiencing a popular boom in local currencies. They are actually being used in about 120 to 130 locations nationwide (in 2003). A local currency is normally money that is used within a specific local community. Therefore, in a broad sense, local currency can also include the cash and demand deposits for which regional financial institutions act as a medium and that are used in the regional distribution of funds. However, the narrower meaning of local currency is money, or a substitute token, that a local community issues on its own and which is only valid within that community. It is local currency in this narrow sense that is currently experiencing a popularity boom. Typical examples are "Oumi", issued in Kusatsu City, Shiga Prefecture; "Peanuts", issued in Chiba City, Chiba Prefecture; and "Kurin", issued in Kuriyama-machi, Hokkaido.

The ways of using local currency are diverse depending on the objective of its use. There is currency that can be used in overall transactions of goods and services together with money. There is currency that is separate from the economic market and limited to use in exchange for homemade products, or used products, or for mutual aid. There is also currency used as payment for the "price" of volunteer activities. The currencies also come in diverse forms – such as paper coupons, minted coins, account books and electronic money. Generally speaking, the local currency systems are small in scale, with the participants per group numbering from several dozen to a hundred or so people.[8]

According to Michael Linton, who established LETS (Local Exchange Trading Systems), it is possible to change the shape of a local economy by limiting the currency's scope of circulation to the community concerned, and furthermore by excluding objectives other than circulation, such as savings and loans (Maruyama 1988). First of all, the exchange of goods and services within a local community will become brisk as a result of the currency constantly circulating within the community. In addition, it will break down the structure separating the two main economic entities, producers and consumers, and create a situation in which any local resident can be both a producer and a consumer at the same time. It will encourage the development of individual abilities as producers as a result of people discovering the needs of their neighbours through communication between residents. Furthermore, one can even involve the government and the landowning class in the system through the use of local currency to pay taxes and land rent. Their involvement would make it possible to boost local welfare services and promote land use that is in harmony with the local environment. Most of all, it would make it possible to grant societal recognition to human activities that conventionally have not been regarded as producing value.

The objective of the LETS type of system is to revive mutually compensatory human relations within local communities, something that was destroyed by the establishment of the market economy. However, it should be noted that since the basic structure of the agrarian society has collapsed, one can no longer have a community that is self-sufficient and closed. For the time being, the idea is to use as the premise the production of goods and their consumption and to change the flow of goods to a regional circulation type system by expanding mutually compensatory relations, and, along with this, invigorate the exchange of non-market-based goods and services.

The reasonable scale of participation in the use of a local currency at the local level would probably be from several dozen to several hundred people. If a system exceeds that, mutually compensatory relations will become weakened and the currency's function will decline. Therefore, if a system grows too large, there are cases where it would be desirable to divide the original system into a multiple number of smaller systems. At any rate, one can anticipate that the distribution of several local currencies within a city will uncover diverse needs within the region and broaden its citizens' realm of activity as producers.

As for the regional economies at the prefectural level, it is reasonable to propose a regional currency to establish a circulation system that links a city with its neighbouring towns and villages. Originally, the clan-issued paper money of the Edo period, mentioned in the previous section, was a regional currency of that nature. Once again, the example of Okutama cedar will provide us with a concrete picture of how a regional currency would contribute to the revival of the use of local and regional resources. Suppose an NPO (non-profit organization) is responsible for issuing the regional currency and gives it a name such as "Tokyo Cedar" or "T-Cedar". The currency would be backed by the actual cedar in the Okutama area, and the NPO would issue "T-Cedar" to the volunteers who would go to the Okutama area and assist the forest owners to take care of the trees. The NPO could be financially supported by any organization and individual who would be interested in the use-value of Okutama cedars. These supporters would be willing to accumulate "T-Cedar" from the volunteers by offering various goods and services which the volunteers could accept. In this manner, the supporters would get cedars in exchange for "T-Cedar". A regional currency could also be backed by a range of regional resources. In this case, the people who accumulate the currency could choose any item in the range.

Regional self-sufficiency and urban sustainability

Without doubt a city depends on its surrounding social and natural environment, and even if it makes a serious effort to reduce its own ecological footprint, the city still has to consume ecological resources of other places. The city needs to be embedded in a regional unity together with surrounding towns and villages at prefectural level in order to reduce the imports from other countries and to make better use of the local and regional resources which were abandoned during the industrialization period. The localization and regionalization of resources means the change of industrial structures, a change which puts more weight on the organic mode of production and gets rid of the unnecessary, and now even harmful, emphasis on the mechanical mode of production. Since the global market system has developed in accordance with the mechanization of human economy, it has adopted a built-in bias towards the mechanical mode of production.

My proposal therefore is to regain an ecologically benign balance between the organic and the mechanical modes of production. In order to promote such a proposal, it is necessary to expand non-market type transactions along with regional market transactions, and both transactions must be embedded in the decentralized regional society. I would not advocate that a regional society should close its borders and pursue only the self-sufficiency route. I only suggest that an increase in self-sufficiency at the prefectural level would not only help to reduce the regional ecological footprint but also broaden the possibility of producing regional wealth from regional resources. Regional wealth consists not only of properties and stocks to be bought and sold in markets but also includes the social capital which enriches people's freedom in terms of "being" and "doing", that is, the freedom to pursue a congenial life-style in a regional society. I strongly believe that it is possible for urban sustainability ultimately to be maintained in a regional society of this type.

Notes

1. According to Book 1 of Aristotle's *Politics*, even when complete self-sufficiency was not possible, the local community should not waver from the principle of self-sufficiency, but should carry out trade only to the extent of supplementing self-sufficiency and should not bring commercial activity into the local community in a disorderly manner (McKeon 1941: 1137–1141).
2. The application of the concept of entropy to matter was popularized by Georgescu-Roegen (Georgescu-Roegen 1971).
3. The question is crucial not only for Japan but also for all other countries which export their goods to Japan. Since Japan's ecological capacity in 1999 was only 0.77 hectares, it will continue to owe other countries including developing ones in terms of ecological footprint even if it reduces its ecological footprint to the world's fair allocation of area per capita which was 1.99 hectares in 1999. This means that although it must continue its trade with other countries even Japan would succeed in reducing its ecological footprint to the level of the world's fair allocation of area. At the international level, Japan must

therefore take the initiative in globally introducing the environmental tax, especially the carbon tax. In this regard, Tobin tax would be also effective as well. The initiative of this kind will surely make Japan take better care of its own environment as well as other countries' ecosystems.

4. "Factor Four" refers to the increase of wealth by 200 per cent while reducing the consumption of resources by 50 per cent. This term was first promoted by the Club of Rome in 1991, and elaborated in Weizsäcker et al. (1997). "Factor Ten" refers to the reduction of the throughput, which is the energy and matter used in the production process, by 80–90 per cent within 50 years, in order to maintain the living standard of 1990. This idea was proposed by the Factor Ten Club in 1995.

5. I would not claim that China and India should be allowed to disregard the world's effort of reducing the burden on Earth. To date, industrialized countries have proposed and undertaken various concrete efforts as environmental cooperation that they should provide to developing countries. These include the transfer of technology and financial assistance for measures to prevent pollution, the utilization of the system for trading greenhouse gas emission quotas, assistance for the afforestation of deserts and actively importing goods from developing countries. Furthermore, not a few of these efforts are producing improved results. Nevertheless, with the exception of a few countries such as Sweden, which is actively promoting the conversion to biomass energy, most countries continue to rely on oil and atomic (nuclear) energy. I would therefore suggest that, if the nations such as the US and Japan become more serious for decreasing their own ecological footprints, China and India would be able to have more freedom of industrializing their nations within the limits of the world's fair allocation of area per capita.

6. The distinction between the organic and the mechanical modes of production is based on Eduard David's study of the nature of agriculture in the socialist economic system (David 1922).

7. The argument regarding the revival of regionalism is mostly based on Ash Amin and Nigel Thrift (Amin and Thrift 1994), Philip Cooke and Kevin Morgan (Cooke and Morgan 1998) and Allen Scott (Scott 1998). The examples of the Basque region and Wales are quoted from Cooke and Morgan.

8. Even large groups include only several hundred to about two thousand participants. Although Argentina's alternative currency system called the RGT is abnormally large with several tens of thousands of participants, the behind this is the situation of the extremely low level of trust in the national currency, the peso, itself.

REFERENCES

Amin, A. and Thrift, N. (eds) (1994) *Globalization, Institutions and Regional Development in Europe*, Oxford: Oxford University Press.

Cooke, P. and Morgan, K. (1998) *The Associational Economy*, Oxford: Oxford University Press.

David, E. (1922) *Sozialismus und Landwirtschaft*, Leipzig: Verlag von Quelle & Meyer.

Factor 10 Club (1995) *Carnoules Declaration*, Carnoules.

Georgescu-Roegen, N. (1971) *The Entropy Law and the Economic Process*, Cambridge: Harvard University Press.

Illich, Ivan (1973) *Tools for Conviviality*, New York and London: Harper & Row.

List, G.F. (1841) *Das nationale System der politischen Ökonomie*, Stuttgart and Tübingen: J.C. Cotta.

—— (1842) *Die Ackerverfassung, die Zweigwirtschaft und die Auswanderung*, Stuttgart and Tübingen: J.C. Cotta.

McKeon, R. (ed.) (1941) *The Basic Works of Aristotle*, New York: Random House.

Maruyama, Makoto (1988) "Local Currency as a Convivial Tool", in *Journal of International Studies*, 3, Tokyo: Meiji Gakuin University.

—— (1999) "Local Currencies in Pre-Industrial Japan", in E. Gilbert and E. Helleiner (eds) *Nation-States and Money*, London and New York: Routledge.

Polanyi, Karl [1944] (1957) *The Great Transformation*, Boston: Beacon Press.

—— (1977) *The Livelihood of Man*, New York: Academic Press.

Sachs, W., Loske, R., Linz, M. et al. (1998) *Greening the North*, London and New York: Zed Books.

Schrödinger, E. (1967) *What is Life? Mind and Matter*, Cambridge: Cambridge University Press.

Scott, A.J. (1998) *Regions and the World Economy*, Oxford: Oxford University Press.

Smith, Adam (1979) *The Wealth of Nations* (Cannan Edition), Tokyo: Tuttle.

Tamanoi, Yoshiro (1978) *Ekonomii to eckorogii (Economy and Ecology)*, Tokyo: Misuzu Shobo (text in Japanese).

—— (1982) *Seimeikei no keizaigaku (The Economy as Based on a Living System)*, Tokyo: Shinhyoron (text in Japanese).

Wackernagel, M. and Rees, W. (1996) *Our Ecological Footprint*, Gabriola Island: New Society Publishers.

Wada, Yoshihiko (2001) "Chikyu no kankyo shuuyou nouryoku to keizai no sai-teki kibo (The World's Environmental Capacity and the Optimum Economic Scale)", in *Jinko to kaihatsu (Population and Development)* 75, Tokyo: Ajia jin-kou kaihatsu kyoukai (Asian Population and Development Association) (text in Japanese).

Worldwide Fund for Nature (2002) *Living Planet Report 2002*, WWF International.

Yano Tsuneta Kinenkai (ed.) (1995) *Sekai kokusei zue (World Census Diagram)*, 1995/1996 Edition, Tokyo: Kokuseisha (text in Japanese).

—— (2002) *Sekai kokusei zue (World Census Diagram)*, 2002/2003 Edition, Tokyo: Yano Tsuneta Kinenkai (text in Japanese).

Weizsäcker, E.U. von, Amory B. Lovins and L. Hunter Lovins (1997) *Factor Four: Doubling Wealth – Halving Resource Use*, London.

4

Land, waste and pollution: Challenging history in creating a sustainable Tokyo Metropolis

Tokue Shibata

1. Conditions for a sustainable society: Japan, Tokyo and post-war economic development

During World War II vast tracts of Japanese urban areas were razed to the ground. Particularly hard hit were the atomic bomb targets of Hiroshima and Nagasaki. There were a few exceptions such as the ancient capitals of Kyoto and Nara, but much of Tokyo was destroyed during the course of more than 50 air raids on the city (one, on 10 March 1945, killed over 100,000 people).

The survivors, many of whom had lost their houses, had to endure great hardship during the early post-war years following Japan's surrender in 1945. The population of the Tokyo Metropolis, for example, which had peaked at 7.35 million in 1940, had fallen to 3.49 million by 1945.[1] Food and fuel were scarce. Boosting food and coal production (under the slogan "Dig as much of nature's black diamond (coal) as much as possible!") were the first goals for survival.

The slogan "increase production and productivity" continued into the 1960s when Japan entered a period of economic growth. People worked hard and endured self-sacrifice believing that their actions would create a better future. As a result, by the mid-1960s, most Japanese were able to afford desirable consumer goods. The automobile industry, an indicator of economic development, grew. Vehicle production exceeded that of individual European countries by 1970 (5.29 million in France and

2.1 million in the UK). By 1980, Japan was producing over 10 million vehicles annually, surpassing the US automobile production figures.

The high-growth period coincided with the rapid urbanization of Tokyo. The capital's population tripled in the 15 years after the war, passing the 10-million-mark in 1961. Concentration and suburbanization progressed. Tokyo became a giant megalopolis, swallowing up the surrounding regions. Today (2005) Tokyo Metropolis has a population of some 12 million, and the Tokyo metropolitan region encompasses over 30 million people. In other words, a quarter of the nation's population lives in or near Tokyo.

No one in the pre-war days or during the war could have imagined the period of mass-production and mass-consumption. But there was another side to this period of post-war affluence: the wasting of precious resources and the generation of garbage. Economists realized that air and water, until then considered infinite resources, could cause great economic losses if polluted sufficiently to affect health and endanger crops. For the first time, people understood that natural resources (fossil fuels and trees) had their limits. They saw that the new "good-equals-rich" lifestyle was imposing a heavy burden on the environment. Too many cars on the roads caused serious traffic congestion, especially in Tokyo, costing people and businesses time and money.

There is a great need for a shift to more environment- and resource-conscious economic activities, with greater emphasis on the 3Rs of "reduce (reduction of generated waste)", "reuse", and "recycle". The "Basic Law for the Promotion of the Formation of Recycling-Oriented Society" was enacted in June 2000, and prompted further laws to facilitate the 3Rs. For example, laws were brought in to cover: product packaging (the "Containers and Packaging Recycling Law"); regulation of electrical appliances including televisions, refrigerators, washing machines and air conditioners (the "Household Electrical Appliance Recycling Law"); food (the "Food Recycling Law"); construction waste (the "Construction Waste Recycling Law") and waste in general (the "Waste Disposal and Public Cleansing Law"; and the "Law for the Promotion of Effective Utilization of Resources"). In January 2005, the "Automobile Recycling Law" was enacted to promote the recovery and recycling of airbags, automobile shredder residue (ASR), and chlorofluorocarbons (CFCs).[2]

For these laws to be effective and reform Japan into a sustainable society, they need to be able to: (1) reduce production of unnecessary goods without hampering free economic activities; (2) guide consumers' purchasing trends and encourage them to buy goods that are easy to reuse or recycle (for example, the US state of Oregon brought in a Bottle Bill in 1971 that introduced a deposit refund system on all drink containers); and (3) minimize the cost and burden of treatment and disposal on the environment. These issues will be discussed as they relate to the case studies examined in this chapter.

In the first sections of this chapter, I describe my experiences with environmental problems in Tokyo from the 1950s to the 1980s. First, I outline the history of modern industrial/environmental pollution in Japan. (Section 2), and then I will review two case studies, the first of which is the "Koto Garbage War" of the early 1970s, when I was working as the director of planning and coordination for the Tokyo Metropolitan Government (Section 3). The second case is that of a citizens' recycling movement in Tokyo's suburban city of Machida that has been active since the mid-1970s (Section 4). These two cases offered valuable lessons to aid us in our efforts to make Tokyo society more sustainable.

In both cases, it was found that land ownership and land use were at the root of Tokyo's pollution and environmental problems, and so I look back to the Edo period (1603–1868) and the Meiji period (1868–1912) to see how land ownership patterns were first established in Tokyo (Section 5), and how the situation changed in the post-war period (Section 6). In the final section of the chapter, I propose some solutions towards the creation of a sustainable future for Tokyo.

2. Outline of modern industrial pollution and environmental destruction in Japan

2.1 Industrial pollution in the Meiji period

Environmental pollution generated by modern industry in Japan dates back to the turn of the twentieth century. The Ashio copper mine pollution case is considered the first example of modern environmental destruction in Japan. The mine (located in Tochigi Prefecture, about 100 km north-west of Tokyo) had been the property of the Tokugawa shogunate, and as such had produced 1,500 tons annually, which was the maximum possible output in the 1600s. However, this high output level had been dropping gradually. The mine was temporarily closed in 1800.[3]

After the Meiji Reformation, entrepreneur Ichibei Furukawa acquired the mine. Using modern techniques imported from abroad, a large new seam of ore was discovered in 1884, and copper production rose rapidly, to reach over 5,000 tons per year.[4] The refining methods at that time, however, were still primitive and generated acidic waste water and industrial waste containing heavy metals such as copper and zinc. These were discharged into the nearby Watarase River where they caused great damage to the farmlands down river. Crops died and the damage spread to livestock and people.

In 1891, Shozo Tanaka, a member of the Lower House from Tochigi Prefecture where the mine was located, demanded that mining at Ashio be stopped. Land tax was either exempted or reduced for the farmers affected. Unfortunately, this meant that the farmers lost the rights given to taxpayers, including civil rights and voting rights enabling participation in the political decision-making processes. As a result of the pollution one village upstream of the Watarase River became uninhabitable, and in 1901, in order to let the public know of the mine poisonings, Shozo Tanaka tried to appeal directly to the Emperor, but his attempt failed. Although Tanaka's appeal did not work out as he had planned, the public at large were astounded by what he had to say.[5] However, the consequences of the pollution lingered on and were not settled for nearly a century.

2.2 Industrial pollution in the 1960s

The boom in economic growth during the 1960s, with its focus on heavy and chemical industries, resulted in severe industrial pollution and environmental destruction. The four major pollution-related diseases of the time were Minamata disease, Niigata Minamata disease, Itai-itai disease and Yokkaichi asthma.

Minamata disease, originating in the city of Minamata (in the southern island of Kyushu along the Yatsushiro Sea), was the greatest pollution scandal to strike Japan in the post-war period. Minamata disease is a neurological disorder caused by mercury poisoning. In this case, it was traced to a methyl mercury compound contained in waste water discharged by the Chisso Minamata plant that manufactured acetaldehyde. The methyl mercury compound in the factory effluent was flushed out into the Yatsushiro Sea, and, through the food chain, the mercury became concentrated in fish and shellfish. Consequently, Minamata disease occurred when the local inhabitants ate large amounts of local seafood. The first patient was reported in 1956 and, by the end of March 2001, a total of 2,265 persons had been certified as victims of the pollution.[6]

In 1965, instances of Minamata disease in the Agano River basin in Niigata Prefecture were reported. The cause was the same as in the original Minamata disease, namely, waste water from the Showa Denko plant that manufactured acetaldehyde. Nearly 700 people were certified as patients by 2001.[7]

Yokkaichi asthma was caused by air pollution from a huge petrochemical complex that discharged sulphur dioxide gas into the air. Itai-itai disease (literally "ouch-ouch" disease, with extremely painful symptoms of renal impairment and bone disease) is chronic cadmium poisoning and was caused by the Kamioka mines discharging cadmium compounds into the local river.

By 1970, the victims of these four pollution cases had filed lawsuits ("The Four Great Pollution Lawsuits") that forced Japanese industry to confront pollution problems. After a quarter-century of passage through the courts, the lawsuits were settled by the mid-1990s, and the surviving victims won all their cases. The government enacted the Water Pollution Law and other laws to impose stricter regulations on industrial waste and waste water. These measures to protect the environment and to ensure more sustainable industrial activities have made dramatic progress, but were only brought into being after great human cost.[8]

2.3 New types of pollution in the 1970s

These four cases of local industrial pollution and environmental destruction all had clear point sources, namely, specific plants or activities. Regional pollution, on the other hand, became a problem during the high economic growth period. Although Tokyo did not experience the kinds of horrendous localized pollution typified by the four cases above, it did have its share of cases, such as falling ash from a cement factory that dated from the late Meiji period to the Taisho period (1910 to early 1920s); the plant subsequently moved to the neighbouring industrial city of Kawasaki. There were also protests by Tokyo fishermen over waste water from a pulp plant in 1958.

In the 1970s new types of pollution and environmental destruction emerged in the processes of mass "production", "distribution", "consumption and waste disposal", coupled with rapid urbanization.

Until the 1970s, the food and daily necessities of the Japanese had, in the main created waste that was generally combustible and biodegradable. Packages and containers were mostly made of wood, bamboo or paper. These could be burned and the remaining ashes buried safely. Any leftovers were either buried in the garden or fed to farm animals. In those days, income was limited, life was simple, and little waste was generated.

As the Japanese population became richer in the 1960s, the quantity of materials used in consumer goods increased rapidly. The ownership of durable goods – including the "three sacred treasures" (namely, black-and-white TVs, electric refrigerators and electric washing machines), heavy furniture, pianos and private cars – increased. As these products reached the end of their serviceable lives (about 10 years for cars and 30 years for pianos) they became discarded as waste. Both the population and per capita amount of solid waste generated rapidly increased during the decade, and in many cities, including Tokyo, this rose to between 10 and 20 per cent annually.

Waste became a particular problem in the rapidly urbanizing Tokyo which lacked a proper system of waste management. Long-suffering Tokyo, which had already lived through industrial air and water pollution, now faced a new threat – that of domestic garbage.

3. The Koto garbage war

3.1 Koto ward: Tokyo's garbage corridor

In the 1960s, waste collected in Tokyo was (and still is) generally transported by trucks through the ward of Koto to the shallow Tokyo Bay, where it was dumped in landfill sites to create "islands" of solid waste. One such "island" is *Yume no shima* (Dream Island). In 1965, a plague of flies originated on the island and spread to the eastern part of the city. The island was also a rodent "paradise", and garbage-fed rats grew bold enough to chase away any cats invading their territories.

Both Tokyo's population and the waste it generated continued to increase rapidly and, as we have already seen, the population tripled in the 15 years following the war, while the amount of solid waste collected rose by between 10 and 20 per cent annually (see Figure 4.1). According to data from the Ministry of Health and Welfare, the estimated per-capita waste generation level increased from 695 grams in 1965 to 921 grams in 1970, or about 30 per cent in 5 years,[9] and waste management became a serious problem for the Tokyo Metropolitan Government (TMG). As two-thirds of the Tokyo garbage landfill was in Koto ward (including *Yume no shima*), Koto citizens believed that they were the victims of noise, odour and other nuisances caused by the garbage brought in by trucks from other parts of Tokyo. They felt that their right to live a clean, healthy life and not be worried by foul-smelling garbage trucks had long been neglected. In response, the local council passed a resolution in 1971 against Koto being a garbage "corridor" and dumping ground for Tokyo's rubbish.

The TMG, particularly the Bureau of Waste Management (responsible for solid waste management), tried to change attitudes in Koto ward. The governor of Tokyo proposed that each ward set up incineration plants for their own garbage.

By this time, several inland wards already had their own incineration plants, and did not rely so heavily on Koto for waste dumping and landfill. One example was the largest ward of Setagaya (its population at that time was around 700,000 – twice as many as in 1945), which had built its

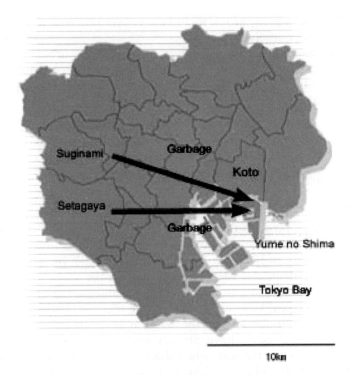

Figure 4.1 Tokyo's 23 wards: major players in the Koto garbage war (*Note*: Waste collected in the 23 wards of Tokyo was generally transported by trucks through the ward of Koto and dumped in landfill sites in Tokyo Bay to create "islands" including Yume no Shima. *Source*: Compiled by author from a map at: http://tokyo23.seisou.or.jp/koujou.htm)

first modern incineration plant (Chitose plant) in 1955, followed by another (Setagaya plant) in 1969. Both these plants used state-of-the-art incineration technology.

On the other hand, Setagaya's neighbouring wealthy ward of Suginami with a population of around 500,000 (which had also more than doubled since 1945), did not have its own plant and completely relied on Koto for its waste disposal and landfill. Koto citizens compared Setagaya and Suginami and criticized Suginami. "Suginami may be an academic and cultured ward", some said, "but it takes no responsibility for the garbage it generates".

In 1971, angry council members of Koto ward passed a resolution to use force to block trucks from Suginami. This led to Tokyo's "great garbage war".

3.2 Construction of Suginami waste incineration plant

Solid waste management was not a high-priority policy issue for the government of Tokyo until the 1960s. The departments in charge of waste management had changed many times since the Meiji Period, from a section responsible for hygiene in the General Affairs Division, to departments in charge of health, welfare, civil affairs, hygiene, waste management, and finally, the environment.[10]

In the pre-war days, Tokyo had one waste treatment facility in Fuka-gawa (Koto ward). In 1939, with the increase in the amount of garbage, the Interior Minister, Koichi Kido, announced a plan to construct facilities at nine sites, including Nishida Town in Suginami. Immediately after this announcement, citizens declared that they "strongly opposed the construction of a nuisance facility", but their protest campaign did not lead to any battles between those supporting the plan and those opposing the plan. Soon after this, Japan entered the Second World War and, under the slogan of "Luxury is an enemy!", the rationing of food and other daily necessities began and the amount of garbage quickly decreased. Hence, the incineration plant construction plan was postponed.

Neglect of waste management and environmental issues continued in Tokyo for a decade after the war until the era when the amount of garbage generated began once more to increase rapidly. The TMG then decided to increase the number of incineration plants based on the principle that "each ward should be responsible for the waste generated in it". (It should be noted that it was believed at that time that burning waste was the best method both of disposal and for protection of the environment. It was not then known that burning garbage could generate dioxin and other toxic substances.)

After the Chitose plant was constructed in the neighbouring ward of Setagaya in 1955, debate on incineration plant construction began in Suginami, and the Waste Management Cooperation Group was formed. At its general meeting of 1959 the decision was reached to build an incineration plant for Suginami, and everyone agreed that a plant should be constructed somewhere in the ward. But the question was "where". Everyone was reluctant to have it in *their* neighbourhood – the "not-in-my-backyard" (NIMBY) syndrome.

In the meantime, the TMG continued to search for a suitable site for the incineration plant. Nishida Town, which had been a candidate before the war, was one possibility, together with an assortment of factory sites, athletic fields, and other open spaces in Suginami.

On 14 November 1966, the TMG announced that the final choice of site for the Suginami plant, which would have a daily capacity of 900 tons, would be an area of about 32,200 square metres in the town of

Takaido, next to the main Loop 8 road.[11] This announcement triggered a strong protest from the local citizens, including the 18 landowners of the site. A signature campaign was organized, and 32,000 people signed a petition that stated that the citizens were against the plan for the following reasons. Firstly, the plant site was in front of the Takaido Station (a private railway) with a concentration of shops and restaurants – and the passage of garbage trucks creating a bad odour and garbage falling from moving trucks would affect both the businesses and the neighbouring residential areas. Secondly, there was an elementary school on the other side of Loop 8, and the passage of nearly 1,000 garbage trucks per day would put pupils at risk. Finally, and most importantly, the TMG selected the site without any discussion with the landowners and neighbourhood citizens. This showed a complete disregard for their human rights and could not be allowed to pass unchallenged.[12]

Koto ward office, on the other hand, criticized the protests in Suginami. Political disputes involving the two wards, their citizens and the TMG continued for the next eight years. Suginami residents were split along two lines: the pro-Takaido plant groups, who would not be affected by having a plant in their neighbourhood, and anti-Takaido groups, who would be affected.

After lengthy negotiations by the TMG, the attitude of Suginami residents began to soften as the TMG tried to win them over by taking the following measures to offset the negative impact of the plant:

(a) Proving that the new incineration plants, unlike the old, dirty plants, would be modern facilities with all possible pollution control technologies.

(b) Promising to construct a tunnel under Loop 8 in response to complaints by residents that garbage trucks passing the local elementary school might endanger children. The design of the garbage trucks was improved to prevent garbage from falling from them and to reduce the smell from their contents.

(c) Using the heat from the incineration plant to supply warm water to an adjacent swimming pool and senior citizens' home, and the construction of a library and other amenities.

3.3 Persuasion of landowners

The most difficult groups to persuade were the landowners. After the war, land prices continued to increase. As Tokyo urbanized, agricultural land was re-zoned to residential land, causing the price of the land to leap by a factor of 10 or more. This created an expectation among people that land was the ultimate asset and should never be sold. It was (and still is) widely believed in Japan that land ownership is an insurance for a safe and secure life. As land is a symbol of its owner's power in the community, it was difficult for the TMG to rely on rational persuasion methods (often successful in the United States, for example) through compensation based on loss of property value, loss of sales, and expense of relocation. On the other hand, the Takaido site owners assumed that the plant would lower the area's land prices and would deprive them of future expected income.

After many twists and turns, an agreement between the TMG and the landlords and residents was reached, as recommended by the Tokyo District Court. By then, over 500 local meetings had been held, creating models for future citizens-vs.-government movements. The plant was constructed at a cost of 17,787 million yen, several times the figure in the original plan, and began operation in 1982.[13] A community centre, swimming pools and senior citizens' welfare centre were constructed to cater for the community.

By the time the Suginami plant was built, however, Tokyo's waste management problem had entered a new phase, and this will be examined in the next section.

4. The recycling movement in Tokyo's Machida City

4.1 From burnable to non-burnable waste

By the mid-1970s, Japan had entered the years of mass production, mass consumption and mass disposal. Consumers now had cars and durable goods. The first instant "cup noodle" product, consisting of dried noodles in a Styrofoam cup, was introduced to the market in 1971. All the consumer had to do was to add boiling water and then wait for three minutes before eating. The big problem was what to do with the cup afterwards. Products like this caused the amount of throwaway plastic refuse to increase rapidly. The government's strategy in earlier decades had been to purchase land and construct large incineration plants, but these faced obstacles as the amount of plastic and other non-burnable waste increased. Incineration of this type of waste would generate toxic gases and residue contaminated with heavy metal and other poisonous substances. Releasing the smoke into the air or burying the residue in landfills would damage the environment on a massive scale.

Recycling now gained attention as a solution to waste reduction, and in this section, I will describe a pioneering recycling movement in the suburb of Tokyo in which I was personally involved.

4.2 The environmental study group in Tokyo's Machida City

Machida is a city in the south-western suburb of Tokyo, about 30 kilo-metres from central Tokyo. It was largely a rural area when, in 1958, one town and three villages were merged and the city created. Machida City's population in 1960 was 71,000.

Many large apartment complexes were built in Machida during the 1960s, and the population rapidly increased from 116,000 in 1965 to 203,000 in 1970 and, by 1981, to over 300,000. In other words, in only two decades Machida's population quadrupled, with many of its residents being commuters to offices in central Tokyo.

This rapid population increase forced Machida City to come up with a new waste management policy for its residents, as by the mid-1970s it was no longer possible to dump the waste in the city's landfill site or burn it on a small scale. To address the issue of waste management, the city es-tablished a study group on environmental issues in 1974, which consisted of scholars, city officials and local residents, and I was invited to chair the group.

The research conducted by the study group over the next few years re-vealed that the economics of recycling waste were quite different from those of conventional products.[14] Normally, when the supply of a prod-uct increases and goes beyond the supply–demand equilibrium point, its price decreases, but the economics of recycling in fact improve with in-creased quantities of aluminium or steel cans, glass bottles, magazines and other waste collected and sorted by each household.

An empty aluminium can left on the roadside has virtually no value. But if cans are collected by the hundred and taken to a recycling dealer, a certain amount of money can be earned, and if a couple of tons of cans are collected, then the scrap metal dealer will be happy to come to the collection centre and get them.

Steel cans are not as valuable as aluminium cans in the recycling mar-ket, but they can be easily separated from other cans by magnets, so the more that are gathered, the greater their value. Old glass bottles and even pieces of broken glass have value, if separated by colour and type, and can be recycled into new glass products over and over again.

4.3 Small houses in Tokyo: No room to store separated garbage

In principle "separation" and "accumulation" are the keys to the suc-cessful recycling of recyclable waste, including aluminium or steel cans, glass bottles, plastic (PET) bottles, and old paper. The more a household separates waste and the more of it they stockpile, the more economically feasible the waste collection becomes. This eventually contributes to maintaining a sustainable society.

The major obstacle to this principle was (and still is) the small area of Tokyo houses. If each house had a 500-square-metre site, there would be enough space for both a garden and a place where separated waste could be stored and organic wastes from the kitchen and garden composted and recycled. However, nearly two-thirds of privately owned housing lots in Machida (that is, approximately 40,000 out of 66,000 housing lots) have an area of less than 200 square metres (as of January 2000). Per-capita area may have been slightly larger a quarter of a century ago, but the specifications of condominiums and apartments have not changed much since then. Beyond the building footprint, houses generally only have a carport and a tiny storage space. Complexes of tiny apartments (one-room apartments) for students and young, single, salaried workers have been built to exploit the maximum legal ground area coverage or maximum floor-area ratio (FAR). The large, company-owned apartment complexes may have enough space for vehicles and bicycles, but not for the storage of recyclable waste.

One promising idea has been to designate collection days, perhaps once a week, when either the city or the contracted dealer collect separated and cleaned aluminium and steel cans, transparent glass, coloured glass, newspaper, fliers and magazines. Each household puts the separated waste either in front of the house or takes it to the collection centre. This system works in households with full-time housewives or retired senior citizens, but is too troublesome for busy working couples having to commute long distances to and from their offices in central Tokyo.

If there were standard specifications for bottles or packages, waste sorting would be a lot easier. Japan traditionally has nationally standardized large 1.8-litre sake bottles used by all sake brands. These are easy to collect and reuse, and are known as the "honour student of recycling" – perfect for a sustainable society. Nearly 90 per cent of sake bottles and beer bottles (the size of the latter is also standardized) are reused. But the Japanese now consume a variety of alcoholic beverages, including whisky, wine and *shochu* spirit (a type of Japanese spirit like vodka), and these bottles come in a tremendous variety of shapes and sizes; moreover, many drinks come in cans or cardboard cartons.

Although the creation of liquor retail shops was strictly controlled until the late 1990s, these shops were not required to have a stock yard of a suitable size for the storage of used bottles for recyclers to collect. Small liquor shops in Tokyo cannot serve as collection centres because they have no room to store large numbers of used bottles.

4.4 Bottlenecks in recycling

The collection of sorted and separated waste in Machida began in the late 1970s on a trial basis. The local authorities specified the collection days according to the class of waste (cans, glass bottles, etc.) and citizens were asked to bring the separated waste on that day.

The major challenge in this new system was to educate residents and gain their cooperation – especially among young office workers who had to leave home early in the morning to commute to offices in downtown Tokyo. The city requested support from neighbourhood and community groups to help these younger people. Machida City's officials and municipal sanitation workers held workshops at apartment complex meeting rooms on evenings and holidays.

Machida City learned from the experience of Numazu City of Shizuoka, where volunteer retired senior citizens were assigned to collection points as watchers to ensure that separated waste was properly collected on the correct day.[15] They were paid a small honorarium either by the city or by the neighbourhood group, and the policy was successful in communities that had these "senior resources".

An aluminium can collection campaign was started in 1974. It was held 54 times until 2004, and over 5 million cans were collected. In addition to these civic movements, collection points to collect plastic waste were designated at supermarkets, civic centres and libraries (72 collection points as of 2003).[16] Thanks to these efforts and public involvement, Machida achieved nearly a 25 per cent waste-recycling rate by the year 2000. This was significantly higher than the national average of 14.3 per cent.[17]

Even though Machida City gradually achieved success in collecting separated, reusable and recyclable waste, another problem emerged – this was, where in the city should the collected waste be stored? Dealers were eager to come and pick up aluminium and other "valuable" waste frequently; however, less desirable waste needed to be collected up for a month or two until the amount was sufficiently attractive to the dealer to warrant collection.

At first, the city used sites in wooded areas and along river banks, arranged through short-term contracts with landowners. However, over the years, urbanization forced the re-zoning of many of these sites into residential land, and cases were discovered where hazardous substances leaching from waste held outdoors had contaminated the soil.

No matter how much citizens understand and cooperate in promoting the 3Rs in an attempt to achieve a sustainable urban lifestyle, the difficulty of setting aside enough land for the storage, treatment and disposal of waste in a highly urbanized environment is a formidable barrier to success. Here again, as in the case of an incineration plant in Suginami, the shortage of suitable land became an obstacle to maintaining urban sustainability.[18]

The next two sections look at the history of land ownership and management in Tokyo in the three periods of Edo (1603–1868), Meiji to the end of WWII (1868–1945), and the post-war period (1945 onwards), and examine the questions of what land means to the Japanese people, why houses in Tokyo are so small, what factors limit the building of more spacious houses with gardens, and the reasons for the lack of urban planning from the perspective of urban sustainability.

5. The history of land ownership and land management in Tokyo

5.1 Land and the Japanese people

Land has historically been intensively utilized in Japan, where a population of some 127 million lives on a total land area of some 378,000 km^2 (the size of the US state of Montana, or a little larger than Finland) of which three-quarters is mountainous.

Throughout its history of over 2,000 years, the Japanese have largely been an agricultural people with a strong attachment to the land, and land has been an important property since ancient times. The *bushi* (samurai), who originated in the late Heian Period (794–1192), were given land (mostly in eastern Japan) by the Imperial Court. Land was granted to those *bushi* who distinguished themselves in battle.

5.2 Land use in Edo

In 1603, after a century of civil war, Ieyasu Tokugawa was appointed Shogun by the Emperor and established his government in Edo, the old name for the city of Tokyo. (By coincidence, the history of another great city, New York, begins around this time when an English explorer Henry Hudson discovered Manhattan Island.) The Tokugawa Shogunate ruled Japan over the next 260 years or so under a highly centralized feudal system. During this period of peace and general stability under a national isolation policy, by 1720 Edo's population had grown to more than a million, making it probably the largest city of that time.

Due to the irregular topography of Edo, with its many hills and valleys, the settlement pattern within the city was complex. About 70 per cent of the land in Edo's built-up areas was owned by the ruling samurai warrior class, including the Tokugawa Shogunate (who lived in Edo Castle), some 300 *daimyo*, and lower-ranking samurai. The *daimyo* were rulers of local districts, which the Shogunate controlled by forcing them to maintain residences both in their own estates and in Edo (typically three in Edo). The *daimyo* were required to travel back and forth between their domains and Edo in order to pledge allegiance and to serve the Shogun, while their families were kept in Edo as virtual hostages. Large *daimyo* estates were typically located on hilltops and hillsides. The valleys were inhabited by lower-ranked samurai, merchants, peasants and tradesmen.

The remaining 30 per cent of the land was divided between shrines and temples, and the commoners (merchants and artisans). The lowland areas near the sea, known as the *shitamachi* (lower city) were heavily populated by commoners. In the early eighteenth century, over half of Edo's population of over one million lived and worked in the *shitamachi* districts which occupied only 16 per cent of the total land area.

Theoretically, samurai land belonged to the shoguns, who bestowed fiefdoms to the *daimyo*, who, in turn, collected dues (rice) or taxes from the farmers in exchange for the right to use the land.

5.3 The Meiji Restoration and land redistribution

With the collapse of the Tokugawa regime in 1868, Edo was renamed "Tokyo" and a new government established. The following year, in 1869, the *daimyo* returned their land registers to the Emperor, and many *daimyo*, together with their samurai, left Edo and returned to their own domains. For the next several years, much *daimyo* land was abandoned. The Tokyo governor encouraged the planting of tea bushes and mulberry trees on these lands, as tea and silk (silkworms eat mulberry leaves) were the major export products of the modernizing Japan. Although well-meaning, the attempt ultimately failed.

The new Meiji government, after abolishing much of the feudal system, proceeded with reforms of property ownership and taxation. Lands came under the new ownership of members of three groups.

5.3.1 The Meiji Government and its officials

In October 1968 the Meiji Emperor entered Edo Castle, renaming it "Tokyo Castle", and became the owner of two ex-*daimyo* estates: Akasaka Detached Palace (now the state guest house for overseas VIPs) and Shinjuku Gyoen (now a public park). The 14 Imperial families were each given residences taken from former *daimyo* property. The area surrounding Tokyo Castle (the Imperial Palace) was used as the site for central government buildings. Many government officials condemned ex-samurai residences, and in some cases forced out the former residents.

5.3.2 The ex-daimyo

During the Meiji period, Tokugawa heirs and those who displayed allegiance to the Emperor were raised to the peerage and thus came to own property in Tokyo. In 1911, there were 11 ex-*daimyo* who each owned at least 30,000 *tsubo* of residential land (1 *tsubo* is equivalent to 3.3 square metres or 35.58 square feet, so this would be equivalent to a land holding of 10 hectares or greater). These 11 *daimyo* between them owned 531,993 *tsubo*, or nearly 1,760 hectares. (The Imperial Palace occupies 115 hectares of central Tokyo.)[19]

This 1,760 hectares or so was not concentrated in one block but was distributed throughout Tokyo. For instance, Morinari Hosokawa, an ex-*daimyo* of Kumamoto, owned 49,000 *tsubo*, and his residence stood on a spacious 30,000-*tsubo* lot in what is now Bunkyo Ward in Tokyo. The remaining 19,000 *tsubo* were divided into 60 lots in nine towns. In another case, Masataka Okouchi, a prominent Meiji engineer and businessman, owned 17,212 *tsubo*, which included most of the residential land in the town of Yanaka Shimizu (now part of Taito Ward) as well as some lots in another part of Tokyo (see Table 4.1). The hilly town of Yanaka Shimizu was divided into 72 blocks and Okouchi used four blocks totalling 5,777 *tsubo* as the site of his residence. The other blocks consisted of small lots on the hillsides and valleys where the streets were winding and narrow.[20]

Besides their residential lots, these ex-*daimyo* owned vast tracts of agricultural land and forests, which were gradually converted into residential land as urbanization progressed. It is estimated that about 12 per cent of the area in Tokyo held by samurai in the late Edo period was occupied by ex-*daimyo* until the end of World War II.[21]

Table 4.1 Scattered housing lots owned by Masataka Okouchi in the town of Yanaka Shimizu, 1911

Lot number	Area *(tsubo)**	Lot number	Area *(tsubo)*
16-2-5	44.50	9-2	49.16
16-2-12	89.77	9-4	55.96
4-1	64.00	10-1	32.30
4-3	13.00	11-1	23.74
6-1	89.25	11-3	36.72
6-3	29.00	12-16	44.11
6-4	48.00		
6-5	9.00		

Source: Tokyo Shi-ku Chosa-kai.
Note: *1 *tsubo* = 3.3 m^2.

5.3.3 *The nouveau-riche merchants and businessmen*

The third category of major Tokyo landowners in the Meiji Period was
the nouveau-rich money lenders and merchants. They speculated on
land during the chaotic and transitional period in the early Meiji, as illustrated in the comment: "Land in Tokyo in those days [early Meiji] was
said to be expensive even at 0.25 yen per *tsubo*, but I heard that a lot
near Nihonbashi was sold at 15,000 yen per *tsubo* in 1919".[22] These spec-
ulators became incredibly wealthy as land prices skyrocketed during
the period from the late nineteenth to early twentieth centuries. Some
were introduced in a book entitled "History of the Rich in Meiji" by
Gennosuke Yokoyama.[23] They included the Mineshima family, who together owned 271 lots throughout Tokyo. Kiyo Mineshima, a business-
woman, owned over 130,000 *tsubo* of residential land in Tokyo in 1913.
She created the Owariya Bank in 1899, established the Mineshima Partnership Company in 1915, donated 500,000 yen (then an enormous sum)
to the government of Tokyo for educational promotion, received the
Order of the Sacred Treasure for her achievements, and died in 1918.
Many of the lots in Mineshima's holdings were registered under the
name of "Ko Mineshima", a relative (Table 4.2). Over 179 residential
lots in 61 chos[24] throughout Tokyo were registered as belonging to Ko
Mineshima.

Table 4.2 Some land lots owned by Ko Mineshima

Town name and lot number	Area *(tsubo)*	Assessed value (¥)
Sanbancho, Kojimachi Ward		
57-1	220	NA
84	543	NA
85	351	NA
(a total of 3 lots in this town)	(1,114)	
Yonbancho, Kojimachi Ward		
1	570	2,157 (with a market value over 1.8 times higher than this)
Kakigaracho 2-chome, Nihonbashi Ward		
1-1	588	11,769
1-5	810	17,911
(a total of 9 lots in this town)	(total 6,437)	(total 139,804)

Source: Tokyo Shi-ku Chosa-kai.

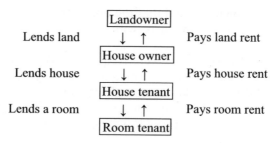

Figure 4.2 The four-layered structure of the rent system (*Source*: created by the author)

5.4 Landowner involvement in real estate

In addition to owning a huge lot for his personal residence, the typical ex-*daimyo* landowner would own a couple of dozen lots scattered throughout the district. The employees or servants of the ex-*daimyo* ran the business of renting out these lots to tenants. (The landowner himself would never meet the lessees). Similarly, the lots the ex-*daimyo* purchased were scattered throughout many towns. These holdings were managed by the landowner's employees or servants, who would lease the land and collect the land or housing rent.

Landlords typically leased their land to house or apartment "owners". These lessees, in turn, built houses and apartments on the leased land, rented them out and collected rent from tenants via their own servants or employees every month. In the case of low-income tenancies, rent was collected every day. Most Tokyo commoners lived in such rented houses, apartments or rooms in the Meiji period. Unused rooms were sub-let to students or single workers, sometimes with meals included (see Figure 4.2).

5.5 Frequent fires and high rents

Meiji's "reforms" did little to help the commoners' housing situation. The low-quality wooden houses of the Edo period gave way to the low-quality wooden houses of the Meiji period. A typical dwelling complex or row house had up to a dozen or so single-room dwellings. Each dwelling was only 9 square metres in area, including the entrance and the kitchen. The high concentration of these apartment complexes allowed the population density in Hashimoto-cho of Kanda Ward to reach 1,516 people per hectare in 1-chome, 822 in 2-chome, and 1,123 in 3-chome in 1873[25] – all amazingly high for a low-rise residential district. A survey of the same ward later in 1884 revealed that the population density had reached 2,857 people per hectare in parts with a high concentration of row houses.[26]

Just as in the Edo Period, fires were frequent in Meiji Tokyo. The congested, wooden, low-quality houses lining narrow alleys were built on the assumption that there would be fire. These low-quality houses were called *yakeya*, meaning "kindling houses".[27] In the first 23 years of the Meiji Period, 121 major fires (fires in which at least 100 houses burned down) broke out in Tokyo.

Although the commoners living in Tokyo during the Edo and Meiji periods in small houses with few personal belongings generated very little burnable and biodegradable waste (human waste was collected, purchased by farmers and used as fertilizers), the houses were not built in a sustainable way. The rising rents for urban houses and apartments were in general not accompanied by improvements in housing quality.

5.6 Ukichi Taguchi's plans for urban land reform: Making Tokyo more sustainable in terms of fire avoidance

In 1882, the economist Ukichi Taguchi released "Fire Prevention Methods" as part of his book *Keizai-saku* (*Economic Policies*).[28] Taguchi was the first person to review Tokyo in terms of "urban sustainability". He concluded that the frequent fires in Tokyo were caused by the concentration of people and goods from all over Japan. He argued that Tokyo in the early 1880s was not a suitable place for modern industrial production, but rather a place where goods and services were wasted by the upper classes (the *daimyo* in Edo and the noblemen in Tokyo) and where amusements like sumo, circuses and stage shows has emerged to entertain their servants and employees who had ample time to waste.[29] Because Edo-Tokyo was a consumption centre and not a production centre, there were no initiatives to accumulate capital and stimulate production or to protect goods and production facilities from fire. The commoners of Tokyo, like those of Edo, had nothing to lose even if their houses burned down; they would simply continue on with the hand-to-mouth existence designed to support the consumers at the top of society.

In this way, Taguchi criticized the lack of urban policies to make Tokyo more sustainable and fireproof. He examined the rental housing business in Tokyo and published an article entitled "It is possible to improve the quality of houses in Tokyo" in *Keizai-saku, Part II* in 1886. He cited apartments for rent in downtown Tokyo of Nihonbashi and Kanda where: "The house owner typically builds a row house for 100 yen, and collects rents totalling 6 yen every month". To the question why profits were so excessive, Taguchi quoted words of a house owner:

"The average life expectancy of a row house in downtown Tokyo is only three years; you have to expect it to be burned down by fire by that time. You must take in 2.8 yen from rent every month just to cover the 3-year depreciation of the 100 yen you have spent to construct the dwelling. In addition, I have to pay land rent to the landowner and house tax; these total 1.2 yen every month. The sum (2.8 and 1.2 yen) totals 4 yen. This means that I can only earn 2 yen of profit a month even if I charge 6 yen of rent. I could reduce the rent to one-third of what it is now only if there were no fires".[30]

The frequent fires caused rents to be outrageously high and forced commoners to rent small, low-quality dwellings. This vicious circle made Tokyo even more vulnerable to fires as its population increased.

To address these problems Taguchi proposed several ideas. First, the use of city planning techniques of zoning and land readjustment to make streets wider, to provide more open space and lower the risk of fires. Second, to require housing owners to construct Western-style rental apartments. Third, to require houses to be more fire-resistant, build five- or six-story apartment buildings to make more efficient use of the land, ban the construction of low-quality row houses, and require apartment owners to build three- or four-story apartments instead.

Taguchi argued that landowners in Western cities typically owned entire blocks and took responsibility for any necessary maintenance or improvements to them. However, as we have seen, Tokyo landowners invariably had small lots scattered throughout the city which were leased to house owners, who in turn built small dwellings and tried to maximize their own profits on the back of short depreciation cycles. There was no room for them to consider the benefits of owning entire blocks. Individuals might make a small profit, but as a whole their loss was considerable. In other words, individuals and the community were not in a "win–win" relationship but a "win–lose" one.

City planning to make streets wider, provide more open spaces, make houses fire- and disaster-resistant, and to make Tokyo structurally sustainable, has continued to lag behind, right up to the present day. Perhaps modern houses, with their supplies of gas and electricity are even more vulnerable to fires and earthquakes than the Meiji row houses of old.

6. Post-war land ownership

6.1 The success of post-war rural land reform

As we have seen earlier, big landowners (government officials, the nobility, and rich merchants) in the Meiji period would normally have one big lot set aside for their residence in addition to many residential land lots scattered throughout Tokyo. The landowner would lease these smaller lots to an operator (house owner) who would then build as many tenant dwellings on the site as possible to maximize the amount of rent he could collect.

Rather than being dispersed, if all the landowner's land was in the form of one or two spacious urban blocks, the landowner might have considered raising the quality of the buildings on the block, which would in turn raise the land value, and subsequently, rental income. In reality, since the land lots were dispersed throughout Tokyo (271 lots in the case of the Mineshima family), the large landowners had no incentive to plan and improve the urban environment.

After Japan's defeat in World War II, various reforms were conducted during the Occupation Period that lasted until 1952. In December 1945, SCAP (Supreme Commander for the Allied Forces) announced a list of reforms. One of these was rural land reform, which was based on a memorandum presented by Wolf Ladejinsky, an agricultural specialist, to General Douglas MacArthur. The Diet passed the Land Reform Bill in October 1946, and Land Commissions were set up in each municipality (village and town) to carry out the reform.

Land owned by absentee landlords was purchased by the government with nominal compensation via the Land Commissions and distributed among the tenant farmers.

Landlords having land in excess of 1 cho (2.45 acres or 99 ares) lost their land. Farmland tenancy was reduced from 60 per cent (part-owners and part-tenants included) to almost zero by 1950.[31]

Tenant farmers, who had suffered from high rents in the pre-war days, became landowners; their morale and productivity increased along with technological innovation. (It should be noted that these small-scale Japanese farmers later faced intense international competition. Rice produced in Japan, for example, is 5 to 10 times more expensive than rice produced in Thailand or Vietnam. Japan's food self-sufficiency rate is now around 40 per cent and the future of the country's food sustainability is under a cloud.)

6.2 The failure of residential land reform

While implementing agricultural land reform, the government drafted a plan to conduct residential land reform in cities. In August 1947, the government (a coalition cabinet formed under Tetsu Katayama, Japan's first socialist prime minister) announced a draft of the Residential Land Law.

The draft stated "One building should be built in one parcel of land . . . The building area of one building should not exceed 2,000 square metres"

(Chapter 1), and specified details of surface rights and the right of lease, to promote utilization of residential land (Chapter 2). This reform was an urban equivalent of the agricultural land reform in rural Japan: to control landlords in cities, to foster private ownership of relatively large houses, and to control land speculation.

Unfortunately, at this point in history, the emerging Cold War between the United States and the Soviet Union, and rising communism in China distracted the policies of the occupation forces away from democratic social and political reform. Both the occupation forces and the Japanese government lost interest in urban land reform, and the draft was scrapped.[32]

One of the aims set out in the draft of the Residential Land Law, namely, to increase the number of private houses, each of which would have a lot area of about 300 square metres, was never achieved in Tokyo. This size of 300 m^2 coincides with the average area of lots provided for the working class in the suburbs of England.

6.3 The post-war housing situation and asset taxes

Though the attempt of urban land reform failed, two main factors contributed to reduce the number of urban landlords: post-war inflation and asset taxes.

During the war and during the period immediately after the war, land and housing rents were controlled by the Rent Control Order of 1939. This order was issued under the National Mobilization Law, forcing the nation to mobilize all people and resources for the war effort. Rents were frozen at the 1939 level, even during the period of inflation after the war (the order was revised several times and abolished in 1961). The pre-war housing rental business was greatly affected by the order, and many houses and apartments for rent, which escaped the bombing, were either used for payment in kind to offset taxes, or sold cheaply to tenants.

The asset tax was imposed only once, in 1946. Many landlords paid the tax in kind (in the form of land) or with cash raised from the sale of land. (Many houses and apartments for rent had been destroyed during the war between 1941 to 1945.) The majority of small and scattered land lot holdings owned by Hosokawa, Okouchi and Mineshima, and other big landlords, were sold, mostly to banks and real estate companies (not to commoners). Later, while land prices soared, these sites were converted to commercial land uses, further subdivided into smaller lots, or used to build large condominiums. Inflation, taxes and rapid urbanization have all contributed to fragmented land ownership in Tokyo of today (see Table 4.3 for price comparison).

Table 4.3 Comparison of price trends among gold, rice, rent, wages and land

(units: yen)

Item	Early Meiji	Circa 1900	1920–25	Post-war (1945–49)	1953–57	1970–72	1990	2000
Gold (1 g)	0.67 (1868)	1.34 (1897)	1.73 (1925)	4.8 (1945) 75 (1947)	585 (1953)	690 (1970)	2,008	1,140
Rice (10 kg)	0.55 (1868)	1.19 (1902)	3.04 (1922)	6 (1945) 149.6 (1947)	845 (1955)	1,600 (1972)	3,741	3,641
Rent (per month)	0.08 (1880)	0.75 (1899)	10 (1924)	50 (1945) 150 (1948)	1,800 (1955)	15,000 (1971)	130,000	110,000
Carpenter's daily wage	0.5 (1868)	0.85 (1903)	3.53 (1923)	35 (1945) 130 (1948)	730 (1955)	3,500 (1970)	12,690	14,790
Commercial land (per tsubo)	5 (1873)	300 (1897)	1,000 (1921)	150,000 ('47) 400,000 ('49)	2.63 M (1957)	6 M (1970)	100 M	40 M

Source: Shukan Asahi (1987).
(Figures for 1990 and 2000 have been obtained via web search.)
Notes:
Gold: the maximum sales price of the year after 1970.
(Figures for 1990 and 2000 from http://gold.tanaka.co.jp/commodity/souba/y-gold.php).
Rice: Price of standard-quality rice.
(Figures for 1990 and 2000 from http://www.shizuoka.info.maff.go.jp/nousei/data/bekasyo.htm).
Rent: Either a detached or semi-detached house with 3 rooms, kitchen and bath in Itabashi, Tokyo (floor area of 40–50 m²).
(Figures for 1990 and 2000 are estimates.)
Commercial land: 1 *tsubo* in Ginza, Tokyo's central commercial district.
(Figures for 1990 – during the economic "bubble" period – and 2000 are estimates).
Daily wage of a skilled carpenter. (*Average of all construction workers for 1990).
(Figures for 1990 and 2000 from the Ministry of Welfare and Labour.)

6.4 Housing lot ownership since 1970s: further subdivision of land

Housing was in short supply for many years after the war. Due to inflation and a drastic decline in the supply of rental housing, the population
relied on the public sector (mostly for rent) and the efforts of individuals
who dreamed of one day owning a house.

Young people who moved to Tokyo during the period of high economic growth and urbanization worked hard, saved money, got married
and then tried to purchase their own homes. They had deposited their
savings (personal savings rates in Japan are much higher than those of
many Western countries) with financial institutions (banks and postal
savings accounts). Unfortunately, the banks at this time focused on lend-
ing money to corporations and industrial activities (such as, purchase of
land, construction of plants and offices, and purchase of machinery and
equipment) rather than to individuals wanting to buy a house or car.
Banks could expect a higher return from corporations, which used purchased land as collateral to borrow even more money from the bank for
business expansion.

Tokyo residents who got married and had two or three children saved
up their deposit, applied for a housing loan from the Housing Loan Corporation (a government agency that provides low-interest and long-term
loans for the construction of houses) and searched for relatively inexpen-
sive land in the Tokyo suburbs. Demand was always greater than supply.
Thus the land price per unit area continued to increase while the land
area that purchasers could afford got smaller and smaller.

Table 4.4 shows the trends of unit land prices, average land area purchased by individuals, and the percentage of land lots of less than 100 m^2
bought by individuals in Koto and Suginami (the two "players" in the
garbage war) and Machida (recycling). As stated earlier, Koto is a downtown ward facing Tokyo Bay, and Suginami is a mostly residential ward
in the western sector of the 23 wards of Tokyo. Machida is further out, in
western Tokyo.[33]

Figures for Koto show the total area of privately held land in 2002
to have been 15,520,000 square metres, of which about 30 per cent
(4,212,000 square metres) was owned by 25,706 individuals. Of these
individual landowners, over 60 per cent (16,224) each owned less than
100 square metres. (This ratio was over 50 per cent in other 10 inner
wards). The total land area owned by those "mini land owners" accounts
for only 22 per cent of all the individually held land. On the other hand,
of the remaining 70 per cent of land owned by corporations, two-thirds
was owned by a small number of corporations each owning over
10,000 square metres. Similar trends can be seen in both Suginami and
Machida.

Table 4.4 Land ownership patterns in Tokyo's two wards and one suburban city (2002)

	Koto	Suginami	Machida	Tokyo Metropolis*
Total area of private land (1,000 m^2)	15,520	22,740	44,100	773,440
Total area of residential land (1,000 m^2)	13,361	20,489	24,730	518,569
• Held by individuals	4,212	17,572	20,987	390,744
• Held by corporations	9,149	2,917	3,743	127,824
Total area of agricultural land** (1,000 m^2)	N/A	127	1,710	16,271
Total number of landowners (private land)	30,335	80,673	75,310	1,788,642
• Individuals (% share)	26,467 (87.2%)	76,805 (95.2%)	73,037 (97.0%)	1,679,729 (93.9%)
• Corporations (% share)	3,868 (12.8%)	3,868 (4.8%)	2,273 (3.0%)	108,913 (6.1%)
Average area of residential land held by each landowner (m^2)	159	229	301	236
Of the individual landowners no. and % of those owning land less than 100 m^2	16,224 (61.3%)	25,537 (33.3%)	6.392 (9.2%)	604,128 (45.7%, 22.4%)***

Source: Created by the author with the data from City Planning Bureau (2002) *Land in Tokyo*, Tokyo Metropolitan Government.
Notes:
*Figures for entire Tokyo (islands excluded) are given as a reference.
**Agricultural land within urbanization promotion (= residential) area.
***The former figure (45.7%) is for the ward area and the latter (22.4%) is for the rest of the area in the suburb.

7. Toward a sustainable Tokyo

To conclude this chapter, I would like to propose some measures towards making Tokyo a more sustainable city.

First, we should put an end to the over-concentration of social, political and economic factors in Tokyo and promote decentralization. Concentration in Edo, and later Tokyo, has progressed over the past 400 years. In the Meiji period the ruling class established a centralized government based in Tokyo in order to modernize the country and catch up with other developed countries. The centralized government system remained after World War II and flourished in the high growth period of the

1960s. Tokyo has become not only the country's political centre but also the centre for the economic, cultural, and all other national functions. The city is covered with buildings, asphalt and concrete, with many high-rise buildings making the maximum use of land. In summer, the air conditioners of these buildings and houses generate heat exhaust, which asphalt and concrete absorb, leading to the "urban heat island" phenomenon. In addition, the ecological sustainability of Tokyo is threatened by its vulnerability to earthquakes.

Decentralization can be achieved by delegating greater political authority to prefectural and municipal governments. As of January 2005 the so-called "trinity reform" (a plan to reallocate central government subsidies and local allocation tax and transfer tax resources to local governments) came under discussion. It is hoped that local governments can take initiatives in implementing policies tailored to local people, make their local areas more attractive in order to entice back their former residents from Tokyo, and invite new residents.

Some people argue that tax reform will only benefit Tokyo as so many corporate headquarters are sited there. However, if a significant amount of indirect taxation (consumption or sales tax, liquor and tobacco tax, gasoline tax and automobile weight tax) is allocated to local governments, the regional imbalance, particularly in terms of per capita amount, will be narrowed.

In addition, inheritance and gift taxes should be reformed. At present revenue from these taxes comes mainly from land, with less from art and cash. The tax rate is high for large landowners (maximum 50 per cent) and a significant amount of this tax is paid in kind in the form of land. The government auctions off this land, but most of it is snapped up by resourceful real estate developers and banks. It would be better for local governments to be given the right of pre-emption. The land the local government purchases in this way (using money from the central government) could be used to stockpile accumulated waste, or other means, to promote a sustainable environment. Property taxes could be changed to reflect new working and living patterns (for example, the increase in the number of very small houses and condominiums, the increase in the number of SOHO workers who work at home or live in their office).

We might once again consider reviving the residential land reform that was left to wither 50 years ago. What kind of "reform" is possible in the Tokyo of the twenty-first century? What policies can we take to stop the trend of residential land lot subdivision and land price increase? Can we somehow limit private land ownership and increase public land? Here, we can learn from the experiences of China who turned all land into public property after the 1949 revolution and is now implementing a socialist market economy.

For over a half century, the Japanese economy placed priority on production and productivity increase. When production increased, so did consumption and waste. The free market gave rise to one fad after another to stimulate consumers and encourage them to use disposable goods or to throw away goods once the products became out of date. The resulting garbage has caused the waste management and recycling problems that we examined earlier. While consumers have been supplied with an abundance of high-quality goods (cars, electrical appliances, computers, cell phones, and so on) most of them still live in cramped houses in congested Tokyo.

We should shift away from mass-production, mass-consumption and mass-disposal, and establish mechanisms for sustainable development of society. We should review our lifestyle and put more emphasis on the 3Rs in our daily life. Ultimately, we should reexamine land policies, which are at the root of urban policies, from various perspectives, including national development strategies, decentralization, financial and fiscal systems, housing and building, city planning and welfare.[34] If we can find solutions to maintain urban sustainability within our existing limited land resources, we can use these to set an example for other cities in similar situations to follow.

Notes

1. Tokyo Metropolitan Government, "Jinko no Ugoki" (Population Trends), 1940–1945.
2. The official title of the law is "The Law on Recycling of Used Vehicles". The law text in Japanese is at http://law.e-gov.go.jp/htmldata/H14/H14HO087.html.
3. Jun Ui (ed.) (1992) *Industrial Pollution in Japan*, United Nations University Press, http://www.unu.edu/unupress/unupbooks/uu35ie/uu35ie00.htm.
4. Ibid.
5. Ibid.
6. Environmental Health Department, Ministry of the Environment (2002) "Minamata disease: The history and measures", http://www.env.go.jp/en/topic/minamata2002/index.html.
7. Ibid.
8. Ibid.
9. Environmental Policy Bureau, Ministry of the Environment (2002) *Kankyo Tokeishu 2002* (Data on the Environment).
10. Suginami Shoyo Memorial Foundation (1983) *Tokyo Gomi Senso – Takaido Jumin no Kiroku* (Tokyo Garbage War: As Recorded by Takaido Residents).
11. Ibid.
12. Ibid.
13. Ibid.
14. Machida City Waste Management Facility Liaison Council (1984) *Gomi to Bunka – Risaikuru Bunka Toshi heno Chosen* (Waste and Culture: Challenges to be a Recycling-oriented City), Kogyo Tosho Kabushiki Gaisha.

15. Ibid.
16. Mayumi Sawaki, Makiko Tanaka, Noriko Nakamura, Yoko Hayashi "Toshi betsu no Gomi mondai to sono taiou 6 Machida-shi" (Waste Management Issues and Measures by City: Chapter 6: A Case study of Machida City)", http://www.asa.hokkyodai.ac.jp/research/staff/kado/mokuji.htm.
17. Ibid.
18. A rural town of Kamikatsu in Tokushima Prefecture with population about 2,200 announced its aim to achieve "zero waste" by 2020. At present waste generated in this town is classified into 34 categories. The town has so far achieved the recycling rate of 79 per cent. Details at http://www.kamikatsu.jp/kankyo/zero_sengen.htm.
19. Tokyo Shi-ku Chosa-kai (1911) *Tokyo-shi oyobi Setsuozku Gunbu Chiseki Daichou* (Register of Title Deeds of the City of Tokyo and its Suburbs).
20. Ibid.
21. Hiroshi Yamaguchi (ed.) (1987) *Kogai Jutakuchi no Keifu* (History of Suburban Residential Land), Kashima Shuppankai.
22. Hiroyuki Suzuki (1999) *Toshi he* (A Study of Cities), Chuo Koron-sha.
23. Gennosuke Yokoyama (1989) *Meiji Fugo-shi* (History of the Rich in Meiji), Shakai Shisho-sha.
24. A cho is a neighbourhood unit in Japanese cities.
25. A chome is a subdivision of a cho, and is normally numbered as 1-chome, 2-chome,... and so on.
26. Yorifusa Ishida (1987) *Nihon Kindai Toshi no Hyaku-nen* (A 100-year History of Japanese Modern Cities), Jichitai Kenkyu-sha.
27. Shinzo Ogi (1980) *Edo to Tokyo no Aida de* (Between Edo and Tokyo), NHK.
28. Ukichi Taguchi (1882) *Keizai-saku* (Economic Policies), Keizai Zasshi-sha.
29. Ibid.
30. Ukichi Taguchi (1886) *Zoku Keizai-saku* (Economic Policies Part II), Keizai Zasshi-sha.
31. Tokue Shibata (ed.) (1993) *Japan's Public Sector*, University of Tokyo Press.
32. Keino Omoto, *Sengo Kaikaku to Toshi Seisaku* (Postwar Reforms and Urban Policies), Nihon Hyoron-sha, 2000.
33. Tokyo consists of 23 wards (which comprise the core of the capital) and cities, towns and villages either to the west of the 23 wards or on islands to the south of Tokyo.
34. Land-related laws in need of review include: Basic Land Act, Comprehensive National Land Development Law, National Land Utilization Planning Law, City Planning Law, Building Standard Law, Land Readjustment Law and Urban Renewal Law.

REFERENCES

City Planning Bureau, Tokyo Metropolitan Government, *Tokyo no Tochi* (Land in Tokyo), (1974–2002) (text in Japanese).
Kankyo to Kogai Editorial Committee (2001) *Kankyo to Kogai* (Environment and Pollution), Iwanami Shoten (text in Japanese).
National Land Agency (from 2000, Ministry of Land, Infrastructure and Transport), *Kokudo Riyo Hakusho* (White Paper on National Land Use), 1975–2000 (text in Japanese).
Shukan Asahi (ed.) (1987) *Nedan no Fuzokushi* (The History of Prices), Asahi Shimbun-sha.

[Tokyo] City Planning Bureau (2002) *Land in Tokyo*, Tokyo Metropolitan Government.

Tokue Shibata (1976) *Gendai Toshi-ron Daini-han* (Study of Modern Cities, Second edition), University of Tokyo Press (text in Japanese).

—— (1985) *Toshi to Ningen* (City and Man), University of Tokyo Press (text in Japanese).

—— (ed.) (1993) *Japan's Public Sector*, University of Tokyo Press.

5

Buildings/city patterns and energy consumption[1]

Takashi Kawanaka

1. Purpose of the study

This chapter focuses on two aspects of urban "sustainability": firstly, how buildings and the city (as an accumulation of buildings) are related to energy consumption, and, secondly, what we expect from the operation of city planning systems to increase the efficiency of energy consumption.

Of all the problems associated with global environmental changes, global warming, due to the burning of fossil fuels and the subsequent increase in the concentration of CO_2 and other greenhouse gases in the atmosphere, is a major concern. The efficient use of fossil fuels and the reduction of the amount of CO_2 generated are of particular importance.

The Kyoto Protocol, adopted at the Conference of Parties III (COP3) in 1997, has set a quota for Japan to reduce emissions of CO_2 by 6 per cent from 1990 to between 2008 and 2010. Although Japan ratified the protocol, CO_2 emissions from both households and transportation are rising. As a result, the notion of "saving energy", which has been a major focus since the first oil crisis in 1973, is gaining importance this time from a different perspective: as a means to mitigate global warming.

Innovation through mechanical technologies (for example, changing the automobile gasoline engine to a hybrid system, and then to a wholly electric system powered by a fuel cell) has long been emphasized as a means to save energy. This chapter proposes that in addition to such "separate" energy-saving measures, we adopt "area-wide" measures to reduce the energy needs of an entire city.

In the following sections some of the "area-wide" solutions are introduced. These are based on the research projects carried out by the then Building Research Institute ("Research Group of Urban Structure and Energy" chaired by Professor Takeshi Koshizuka of the University of Tsukuba) as a part of the "Research and Development of New Technologies in Building and Planning Field for Sustainable Society" project conducted by the Ministry of Construction (now the Ministry of Land, Infrastructure and Transport) from FY1991 until FY1995. The Research Group was established to propose formulae and city planning guidelines to guide the formation of an "energy- and resource-conscious city". The final part of the chapter examines the application of city planning systems and future prospects.

These studies were conducted from the perspective of saving energy, and thus comprise only a portion of the broad theme of "sustainability" in this book. It should also be noted that the studies were conducted more than a decade ago, under the influence of values prevalent during the economic bubble in Japan in the mid-1980s.

2. "Formulae" for resource- and energy-conscious urban planning

In the Japanese board games of *shohgi* and *go*, there exists the concept of *johseki*. A set of *johseki* (or the established tactics set moves, and so on) illustrates the best strategy for two competing players.

Tactics are required, for example, in structuring a city to minimize its resource and energy consumption from the time of its construction on to its management and maintenance. We would like to show these tactics by quantitative analysis under a premise that they be as simple as possible, for resource- and energy-conscious urban planning. Naturally, *johseki* should only be used in *shohgi* and *go* games where appropriate; one cannot always win by *johseki* alone. In this chapter, the word "formula", or the key concepts, will be used instead of *johseki* to propose issues for consideration in operating city planning systems and in implementing urban development projects proposed by the private sector.

Table 5.1 lists 31 sets of formulae as outlined below. Formulae 3 to 5 indicate that, based on the studies of Takeshi Koshizuka (1994a) and Takeshi Koshizuka (1994b), neither high nor low road densities will lead to reduced transportation time or reduced energy consumption. There should be an optimum road density somewhere in the middle (though "congestion" is an uncertain factor that makes further refinement of the formulae difficult).

Formulae 6 to 9 are derived from quantitative analysis of energy used

Table 5.1 Thirty-one "formulae" for energy-conscious urban planning

A. "Formulae" for Basic Urban Planning

(1) Formulae for horizontal development and improvement in urban areas
1. The total area of small, useless vacant space increases as the density of buildings increases.
2. The length of the supply network depends on both the quantity of building units and the area of network coverage.
3. Congestion of a central area surrounded by a radial and loop road pattern is heavier than that surrounded by a grid pattern.
4. There exists a certain road density that may minimize required transportation time.
5. There exists a certain road density that may minimize required transportation energy.
(2) Formulae for energy saving in transportation for commuting and business purposes
6. Promote a modal shift to mass transit that may reduce energy consumption.
7. Promote energy saving in commuting by encouraging people to live near their place of work.
8. Encourage offices to locate according to the traffic they generate.
9. Develop mass transit within a commuting area of each business core to construct multicentric business districts.
(3) Formulae for high-rise buildings for the development/improvement in urban areas
10. There exists a base-to-height ratio, for a model of a city as three dimensional space, that may minimize transportation time.
11. There exists a base-to-height ratio, for a model of a city as three dimensional space, that may minimize transportation energy.
12. Restrain the urban sprawl and utilize unused land and promote multi-storeyed buildings to guide the effective use of land and urban growth.
13. Utilize aerial and underground space in areas with high land use potential to reduce long-distance transportation demand.

B. "Formulae" for Plan Making and Improvement of Urban Areas

(1) "Formulae" for the construction of a district heating and cooling system
14. Consider the scale of investment in district heating and cooling systems to achieve energy-efficient, economically viable operation.
15. Consider the area of the district, the floor area ratio and layout of blocks to determine the optimum coverage of district heating and cooling systems.
16. Heat storage tanks enhance the efficiency of district heating and cooling systems.
(2) "Formulae" for the construction of a co-generation system
17. Save energy and stabilize its supply by introduction of co-generation systems in built-up areas.
18. Save energy and promote economical operation of co-generation systems by planning them to supply energy to buildings that have different energy consumption patterns.

Table 5.1 (cont.)

(3) "Formulae" for utilization of untapped energy sources
  19. There are various temperature levels of untapped energy, therefore consider their use method according to each temperature level.
  20. When utilizing untapped energy, the distance between the heat source and end user should be as short as possible.
  21. Utilize untapped energy from water in sewerage and rivers but with conditions attached.
(4) "Formulae" for urban vegetation
  22. A network of green spaces in urban areas.
  23. Promote the planting of tall trees in each lot.
  24. Promote roof planting.
  25. Promote the covering of building walls with vegetation.
(5) "Formulae" for construction in urban areas
  26. Utilize the advantages of accumulation in urban areas.
  27. Larger buildings save more energy.
  28. Consume less cement and steel.
  29. Maximize the service life of buildings.
  30. Assess life cycle CO_2.
  31. Recover and dispose chlorofluorocarbons (CFCs) at a steady rate.

Source: Created by the author.

for commuting in large cities, based on studies by Tsutomu Suzuki (1994) and Tsutomu Suzuki and Naoto Tagashira (1997). Various proposals have been made on whether to concentrate or disperse the places where people work and live. The formulae suggest that the dispersal of workplaces may not increase the efficiency of transportation energy consumption. Measures should be established to create adequate combinations of business and residential areas that are not too far from each other and linked by mass transit systems. The Tokyo Metropolitan Region, on the other hand, has a high concentration of businesses in the city centre (or the extended centre) and has developed residential areas far away in the suburbs. This has resulted in an enormous amount of energy being consumed for commuting. The environmental burden may be reduced by changing this structure – for example, by relocating businesses, creating adequate and compact business areas, and linking them with residential areas via energy-saving mass transit systems.

Formulae 10 to 13 are based on discussions by Takeshi Koshizuka (1995) and Tsutomu Suzuki (1993) under the assumption that the city is a finite three-dimensional space. These discussions concerned the nature and significance of high-rise buildings for the development and improvement in urban areas. The vertical limit to construction is determined by the base-to-height ratio of energy (or time) required for travel per unit distance.[2]

In city planning, the unplanned and sporadic expansion of built-up areas is called a "sprawl", and is considered undesirable. Formulae 6 to 13 indicate the idea that to save energy, unregulated low-density sprawls should be avoided, and high-density land use be encouraged when necessary.

Formulae 14 to 21 concern the area-wide efficient use of energy in highly concentrated built-up areas (or areas where the accumulation of functions should be guided). Some people argue that high-density accumulation should be avoided and low-density land use should be promoted to reduce the burden on the environment. This argument has two flaws: a) it is often based on the implicit or explicit assumption that total human activity in the city will be reduced; and b) the appeal for lower density and recovery of more natural use of land surface without the support of quantitative studies. If lower-density built-up areas are promoted while maintaining present levels of urban activities, the area of metropolitan regions will expand, energy consumption for travel will increase, and the efficiency of any area-wide energy policies will be reduced. In the study, we took the position that the high-level accumulation of vital urban functions should be guided to avoid low-density metropolitan sprawl, and policies to enable efficient use of energy in the accumulated area should be introduced. As Tohru Ichikawa (1997) states, district heating and cooling using co-generation systems can save energy in built-up areas where demand for electricity and heating (energy for air conditioning included) is high and where there are mixed consumption patterns. To make co-generation systems profitable, the step-by-step investment into energy infrastructure in line with long-term high-density development is necessary. The formulae give several hints of how these purposes can be achieved.

Formulae 22 to 25 concern the use of greenery in cities. Urban green spaces suppress the rise of urban air temperatures in summer (that is, the heat island phenomenon), and indirectly reduce the need for air conditioning. Policies to promote rooftop planting have already been introduced.

Formulae 26 to 31 deal with issues that were present at the time the study was conducted, but which are no longer so urgent. In some cases, outcomes of recent research have been utilized or international consensus has been achieved.

I think a set of formulae is effective and suited to the narrow plains and densely populated cities in Japan.

3. Area-wide energy-saving measures

The research project carried out by the Building Research Institute developed a method to determine areas in the Tokyo Metropolitan Region

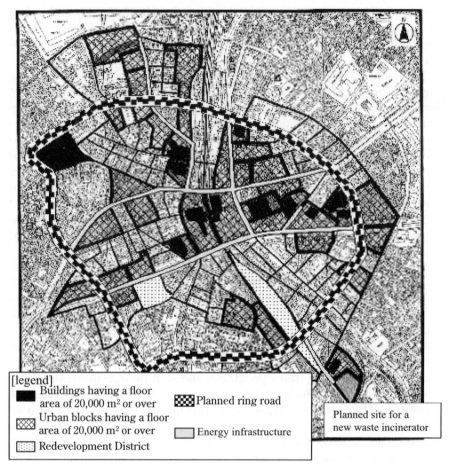

[legend]
Buildings having a floor area of 20,000 m² or over
Urban blocks having a floor area of 20,000 m² or over
Redevelopment District
Planned ring road
Energy infrastructure
Planned site for a new waste incinerator

Figure 5.1 A scenario of energy-saving measures in the Shibuya Station District (*Source*: created by the author)

where area-wide energy infrastructure improvement (as shown in Formulae 14 to 21) might be desirable. A case study was conducted in areas that were suitable candidates for the introduction of district heating and cooling systems. Various scenarios of energy-saving measures were prepared, and the energy-saving effects of each scenario evaluated. The outline of concepts presented in Kazunari Watanabe et al. (1997a) and Kazunari Watanabe and Takashi Kawanaka (1997b) is as follows.

First, mesh data on land use containing numerical information (i.e. Detailed Digital Information (10 m Grid Land Use) provided by the Geo-

Assumed scenario		Estimate results	
Scenario No.	Content of the scenario	Amount of energy saved (Tcal/y) 50 100 150 200 250	Rate of energy saved (%) 5 10 15 20
Scenario 1	Introduce CGS to a building having a floor area of 20,000 m² or over, and supply heat to the rest of the block	(bar)	(bar)
Scenario 2	In addition to Scenario 1, supply heat to other blocks that have a total building floor area of 20,000 m² or over	(bar)	(bar)
Scenario 3	In addition to Scenario 2, supply waste heat from the incineration plant to blocks via a thermal conduit	(bar)	(bar)
Scenario 4	In addition to Scenario 2, use waste heat from the incineration plant to generate power Scenario 5	(bar)	(bar)
Scenario 5	In addition to Scenario 2, use waste heat from the incineration plant to operate a "super" waste-fueled power plant	(bar)	(bar)

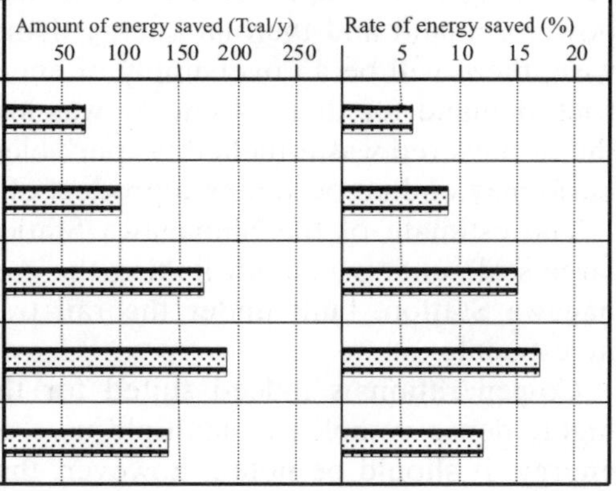

Notes:
CGS: Co-generation system
Block: classified by block number assigned to addresses

Figure 5.2 Estimate of five energy-saving scenarios in the Shibuya Station District (*Source*: created by the author)

graphical Survey Institute) about the metropolitan region was processed through 100 m mesh to estimate the floor area by building use, and to isolate a) areas with high concentrations of commercial and business functions and b) areas where commercial/business functions and middle- to high-rise residential buildings are either in close proximity to one another or mixed. These are areas that have potentially high demand for co-generation and other area-wide energy infrastructure facilities. Some areas have already started district heating and cooling services.

The areas identified in the analysis were narrowed down, by excluding areas where large-scale heating and cooling services had already begun or were already planned. As a result, two districts were chosen: a) the area surrounding the Shibuya Station ("Shibuya Station District" – Shibuya is a major sub-centre in Tokyo) and b) the area surrounding Shinagawa Station ("Shinagawa Station District" – this is an area where the development of new intensive and multi-use urban functions is guided). In each area, the effect of service introduction was estimated based on different scenarios.

Figure 5.1 is a conceptual chart with various hypothetical energy-related facilities in the Shibuya Station District.

Figure 5.2 shows the results of an estimate of energy-saving effects based on several scenarios assuming energy-saving facilities will be developed in the area. The estimates indicate that energy infrastructure should be developed or improved based on the accurate estimation of electric power demand and thermal energy demand of the built-up area; otherwise, there will be an oversupply of energy. In the Shibuya Station District, demand for thermal energy was lower than expected. As a result, the re-powered waste-fuelled "super" electric generation plant in the scenario may end up becoming a product of over investment.

The estimate of the Shinagawa Station District has indicated that a large-scale energy conduit linking the east and west sides of the JR Shinagawa Station, built under the rail track, may prove to be an over-investment.

Co-generation is indeed suited for the efficient use of energy in a highly dense complex of multiple uses consuming a massive amount of energy. It should be noted, however, that I do not recommend energy-wasting urban activities.

4. Expectations towards city planning systems and their implementation

There are various land use control systems in Japanese legislative city planning systems. We shall have a look at the current city planning systems from a viewpoint of energy conscious implementation of urban planning.

4.1 Zoning methods

There are several methods that designate land use zones on a large scale, to control both the formation and scale of cities in Japanese city planning systems. These include designation of "city planning area" and "quasi-city planning area". Many "city planning areas" are further divided into urbanization promotion areas and urbanization control areas (delineation or subdivision known as *senbiki*).

As shown in Formula 12, low-density urban sprawl can theoretically be controlled by the following two types of definitive designation: a) an "urbanization promotion area" to limit the scope of urbanization, and b) an "urbanization control area" to control and restrict urban development. Within the latter, development using special permits and variances should be controlled (prohibited in principle). We can expect these systems to curve the increase of burden on the environment.[3]

The Japanese city planning system has a "zoning system" that designates 12 use zones mainly in urbanization promotion areas. The system prescribes permitted building uses and upper limits of floor area ratios in each use zone. Many kinds of use zones allow mixed land use, without limitation of building use. On the other hand, the use zoning system can induce the inner structure of the city to change according to the method of zoning system implementation. It is possible to allow dense land use to some degree at a site with great economic potential for building, according to the combination of floor area ratio arranged for each use zone category. In short, there is room for mixing or approximating commercial-business use and residential use in use zoning, based on Formula 17 and Formula 18. The current zoning system is effective enough to guide both building use and density represented by floor area ratios. Thus we consider that the well-designed implementation of the zoning system – in co-ordination with energy infrastructure construction and transportation infrastructure construction – in the future should save energy.

4.2 Master planning methods

The Japanese legislative city planning system can be interpreted as having two kinds of master plans at the "upper level" that describes basic guidelines and directions of urban planning.

The first kind of master plan is the "Policies on Improvement, Development and Conservation in City Planning Areas", abbreviated as the "City Planning Area Master Plan" (CPA Master Plan). The City Planning Law (the "Law") was revised in 2000 and the former "Guidelines for Improvement, Development or Conservation in Urbanization Promotion Areas and Urbanization Control Areas" was modified. It is now called the "CPA Master Plan". Each prefectural government prepares "CPA Master Plans" by setting a goal for some 20 years in the future. The "Guidelines for the Operation of City Planning Systems" announced in 2000 indicates some of the guidelines related to reduced burden on the environment (shown in Formulae 1 to 25) should be included in the "CPA Master Plan". These include guidelines for land use, urban facilities/urban development projects, improvement and protection of natural environment and development permit system. For example, some of current CPA Maser Plans mention CO_2 absorption effect by green spaces, but they do not refer to global environment problem directly.

Senbiki, or delineation/subdivision, is an important tool for separating "urbanization promotion areas" from "urbanization control areas". The planning of transportation systems is the key tool for shaping patterns of urban facilities to be developed or improved. For the improvement and protection of the natural environment, the "Basic Green Plan" stipulated by municipalities based on the Urban Green Space Conservation Law, can control the location of greenery in cities (Formula 22).

It is possible to provide the "Guideline for Urban Redevelopment" in the city plan when we choose urban redevelopment in a densely built-up area suitable for the introduction of energy-conscious technologies.

Furthermore there are several planning systems called "sectional master plans" that are not based on statute. We may consider them as a part of the Master Plan in the wide sense. Akio Asakura and Teruhiko Yoshimura (1999) examined how sectional master plans, the Basic Environmental Plan, and the City Plan, can be coordinated. They found significance in coordinating them at the municipal level (the "Municipal Master Plan" – to be mentioned later). We think that it is desirable to coordinate them at the prefectural level given the wide scope and influence of the Basic Environmental Plan.

The second group of legislative master plan is the "Basic Policy on City Planning at Municipalities", which is sometimes called the "Municipal Master Plan" or the "Urban (or City) Master Plan". This plan was introduced when the City Planning Law was revised in 1992. A municipality can now formulate its own unique plan. Many municipalities have adopted their own master plan by involving citizens in the formulation process. Its expected roles are to indicate: basic policies for planning; the city's future image; programmes or procedures; the arrangement of measures; coping with new tasks; the rationale and proper place for control and project; and creating opportunities for participation by residents. Environment-conscious planning should be included among them.

The "Municipal Master Plan" should include both an overall concept and local concept on a spatial scale in one municipality. The city structure of density and scales or locations of public traffic facilities that are conscious of environmental burden might be presented in the total concept. The local concept should present a policy for designation of use zones mentioned above, and each future image of a city, as a premise of planning. It also should include a policy for district planning that we will discuss in Section 4.3.

There have been discussions of the double standards and redundancy of the "CPA Master Plan" and the "Municipal Master Plan" since the time when the "CPA Master Plan" was still called the "Guidelines for Improvement, Development or Conservation in Urbanization Promotion Areas and Urbanization Control Areas". I think we should view countermeasures against global environmental problems as a fundamental direction of all administrative implementations in the future, not only as a part of sectional master plans in urban planning. To consider that countermeasures might work better on a larger scale and the energy conscious methods might be effective in built-up areas and suburbs of large- and middle-scale cities, the following points will be important. Firstly, it is necessary to encourage the mitigation of environmental burden by the "CPA Master Plan" prescribed by the prefectural government. Secondly, it is desirable to show a programme of concrete, individual and detailed environmental mitigation measures in the "Municipal Master Plan", created by the participation of local residents. There are already some environment-conscious municipal master plans that include environmental mitigation measures.[4]

4.3 District planning methods

The Japanese statutory urban planning system has a district planning system that can prescribe controls for detailed building use, floor area ratio limits, and so on. On the other hand, the use zoning system permits the selection from a fixed menu of controls nationwide. The district planning system, introduced in 1980, has now a variety of 'menus'.

It is desirable to utilize a district plan to implement the "formulae" for energy-conscious urban planning methods described in this paper within a limited area (a few to dozens of hectares). It is possible to induce highly dense land use, to save green open spaces, to construct local co-generation plants, and so forth.

There was one case of urban redevelopment district plan, which is a kind of district plan, that treated the construction of a district heating and cooling system as a bonus-incentive factor for a floor area ratio. In district planning systems, it is desirable to realize an incentive or obligation mechanism for a project that can be evaluated by the "CPA Master Plan" and the "Municipal Master Plan" as a promotion area for the introduction of energy conscious methods.

4.4 City planning systems at a turning point

In the earlier sections, I outlined Japanese city planning systems and examined how the two goals of reducing the burden on the environment and promoting energy-saving measures within the context of the present city planning systems can be achieved. As examined, environmental problems can be addressed (or should be addressed) by utilizing existing city planning systems properly.

In addition to the city planning systems described in previous sections, several systems for the protection and improvement of the urban environment have been in operation for about a decade. These include the "Basic Environment Plan" and the "Urban Environment Plan".

The emergence of these systems indicates that the world has changed and that environmental mitigation should now be an integral part of any plan – urban or otherwise. To make environmental mitigation measures work, however, they should not only be required in city planning but should also be accompanied by tax incentives, subsidies, and interest subsidies from the public sector to ensure profitability and stable operation. The examination of these issues is beyond the scope of this study.

Now that Japanese society is demanding an end to the prolonged recession, in the operation of city planning systems, "speeding things up" is more important than ever before. Deregulatory measures and schemes enabling the private sector to make recommendations on city planning matters are examples of methods to speed up the planning process. We should be careful, however, to avoid increasing the environmental burden over the medium to long term.

Decentralization is another trend observed in Japan's public administration. When the City Planning Law was amended in 2000, more power was delegated from the national government to prefectural and municipal governments. Municipal Master Plans were given greater roles than ever, and there is now a greater possibility that neighbourhood citizens get involved to raise levels of environmental consciousness and reflect grass-roots planning activities in upper-tier plans.

On the other hand, these "local" measures should be coordinated with national policies (such as "top–down governance" of global-environmental-change measures by the national government based on international agreements). The regional coordination of various city plans is still a difficult task.[5] One reason is that the level of competition among different cities and regions within Japan (and sometimes across national borders) is intensifying. The amended City Planning Law has delegated the authority to perform *senbiki* from the national government to prefectural governments, and thus local governments now have more freedom to guide and control development. Some people express concern that a nearsighted local government may abolish the *senbiki* system and engage in commercial overdevelopment in the suburbs. By the way, a lot of Municipal Master Plans coordinate the connection of artery roads with adjacent plans, and then the number of coordination on *senbiki* may increase.

Notes

1. This chapter is based on the Building Research Institute et al. (1997), Takashi Kawanaka (1997) and Takashi Kawanaka (2000) and was rewritten after the amendment of the City Planning Law.
2. Osamu Kurita's article of 2001 introduces further discussions of the topic and outlines a new simulation study incorporating population growth.
3. Many critics argue that the primary purpose of the *senbiki* system introduced by the 1968 amendment of the City Planning Law was not to control the expansion of built-up areas by delineating urbanization "promotion" areas and "control" areas; but rather, to prepare for future development. The process goes as follows: a) assign an urbanization "control" area; b) find an area suitable for future development within the "control" area, con-

duct land readjustment and infrastructure improvement, and allow development there at a suitable point in time.

4. Kiyonobu Kaido (2001) points out that several municipal governments have formulated environmental and other administrative Municipal Master Plans with environmental mitigation as their major goal, from the viewpoint of making their cities a "compact city". Typical plans are Aomori City Master Plan (URL: http://www.city.aomori.aomori.jp/toshi/mata00.html) and Fukui City Master Plan (URL: http://www.city.fukui.fukui.jp/rekisi/tosi/masterplan/index.html).

5. Takashi Nakamura (2001) expresses concern for municipalities formulating their own unique plans and the lack of their coordination from a regional and comprehensive perspective.

REFERENCES

Akio Asakura and Teruhiko Yoshimura (1999) "Issues of urban master plans from the view point of environmental policy", *City Planning Review* 48(2): 31–34 (text in Japanese).

Tohru Ichikawa (1997) "Possibility of efficient use of co-generation in cities", *Operations Research* 42(1): 20–25 (text in Japanese).

Kiyonobu Kaido (2001) *Compact City*, Gakugei Shuppan-sha (text in Japanese).

Takashi Kawanaka (1997) "Operation of city planning to achieve energy-saving policies", *Operations Research* 42(1): 7–13 (text in Japanese).

―――― (2000) "Formulae for energy conscious urban planning", *Comprehensive Urban Studies* 71: 131–145.

Building Research Institute of the Ministry of Construction, Japan Institute of Construction Engineering (1997) "Guideline for the planning of resource- and energy-saving urban area" (text in Japanese).

Takeshi Koshizuka (1994a) "Road densities and travel time", *Operations Research* 39(5): 237–242 (text in Japanese).

―――― (1994b) "Road densities which bring a minimum car travel time and a minimum petrol consumption", *Papers on City Planning* 29: 319–324 (text in Japanese).

―――― (1995) "Compact proportion of a city with respect to travel time", *Papers on City Planning* 30: 499–504 (text in Japanese).

Osamu Kurita (2001) "Mathematical model of a compact rectangular city", *Japanese Journal of Real Estate Sciences* 15(3): 39–48 (text in Japanese).

Takashi Nakamura (2001) "Compact city and land use planning", *Japanese Journal of Real Estate Sciences* 15(3): 18–24 (text in Japanese).

Tsutomu Suzuki (1993) "A study on optimal 3D urban form from a viewpoint of compactness", *Papers on City Planning* 28: 415–420 (text in Japanese).

―――― (1994) "Effect of reducing energy used for commuting transportation through optimal allocation of workplaces and residential areas", *Operations Research* 39(5): 243–248 (text in Japanese).

Tsutomu Suzuki and Naoto Tagashira (1997) "What is the energy-saving urban structure from the perspective of urban transportation?", *Operations Research* 42(1): 14–19 (text in Japanese).

Kazunari Watanabe, Takashi Kawanaka, Tohru Ichikawa and Yutaka Tonooka (1997a) "Extracting areas to promote energy-saving measures including co-generation", *Operations Research* 42(1): 26–32 (text in Japanese).

Kazunari Watanabe and Takashi Kawanaka (1997b) "A methodological study on the introduction of co-generation systems (CGS) in Tokyo Metropolitan Region for energy-saving", *Proceedings of International Symposium on City Planning: New Paradigm in City Planning*, Nagoya, Japan, pp. 139–148.

6

Car use and sustainability: Reflection on retail development control systems

Kiyoshi Takami

1. Introduction: retail developments and car use

1.1 Suburbanization of retail developments and sustainability

Over the past few decades, the suburbanization of retail developments has progressed in many Japanese cities, alongside an increase in the suburban population and the widespread use of automobiles. Locations of new shopping centres have moved from city centres to suburbs, as shown in Figure 6.1. This has resulted in the devitalization of many traditional shopping areas, ranging from large shopping districts in the city centres to small local shopping areas.

Against this backdrop, the revitalization of those traditional shopping areas has been called for, helped by nostalgia for the "good old days" when they represented the heart and soul of our hometowns. The response of the Japanese government to this call has included infrastructural development and the construction of car parking in traditional shopping areas, as well as regulations on the establishment of large-scale stores under the Large-scale Retail Stores Law. However, such measures have not been enough to halt the decline of urban centres. Poor linkage between commercial policy and urban planning due to the vertically divided administrative structure has been cited as one of the reasons for this failure (Minohara et al. 2000). More recently, three laws related to urban planning and renovation have come into force since 1998, namely,

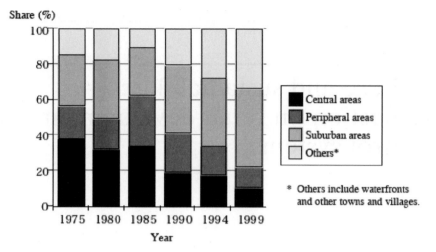

Figure 6.1 Location of new shopping centres (1975–1999 comparisons) (*Sources*: White Paper on Construction, 1995; Japan Council of Shopping Centres press release, 2000)

the revised City Planning Law, the Downtown Revitalization Law and the Large-scale Retail Stores Location Law.

The suburbanization of retail and other developments not only devitalizes existing city centres and shopping areas but also has adverse effects such as increased car use, traffic congestion in the suburbs and loss of suburban natural environment. Thus, suburbanization undermines environmental sustainability.

In addition, the decline of neighbourhood shopping streets and central shopping districts, which are easily accessible on foot, by bicycle or by public transportation, will deprive those without cars of the opportunity to access shopping areas, subsequently affecting social sustainability. Given that the change in location, scale and/or nature of retail developments conforms to economic rationality, however, putting restriction on it might impact negatively on economic sustainability.[1]

1.2 Retail development control in terms of restraining car use

Some countries and cities have adopted policies to regulate retail developments in the suburbs where public transportation service is inadequate, and to enhance and develop "centres" where opportunities for various types of activity, including retailing, are concentrated (Takami 1998). The purpose of such policies includes maintaining and revitalizing centres and reducing the use of automobiles. In a previous study (Takami et al. 1998), I identified the elements that are commonly found in the advanced cases of land use control measures in terms of reducing car use. According to the findings of the study, retail developments should be guided in the following manner:

1. Restrict locations to the urban peripheries;
2. Encourage locations in the vicinity of public transportation nodes (stations and/or bus stops);
3. Strike a balance between various levels of centres, ranging from city-wide centres to small local centres;
4. Ensure a mix of different functions in centres.

The first requirement stems from a principle that population concentrations should have a commensurate retail concentration at their centres in order to minimize travel distances. The second requirement is based on the idea that shorter egress from a node will promote public transportation use, thus reducing the use of automobiles. In many cases, it will be possible to meet these requirements by focusing developments on existing centres within urban areas that are already well served by public transportation.

The purpose of the third requirement is to reduce the number of longer-distance trips by car. This could be done by concentrating comparison shopping stores and major facilities with larger catchment areas at higher-level centres, while locating convenience goods stores and localized facilities with smaller catchment areas at lower-level local centres, Thus providing residents with amenities to meet their day-by-day needs in their neighbourhood. However, the provision of amenities at local centres would have little effect if the choice offered was not attractive enough or if those offered further away (e.g. in higher-level centres or in distant suburbs) were far more attractive. Therefore, an appropriate balance should be ensured between different centres.

The fourth requirement is intended to facilitate multi-purpose journeys through a proper mix of amenities in a centre, subsequently reducing the frequency of trips required to satisfy all the needs. It will also serve to reduce the distance of individual journeys in a trip chain, many of which will be made on foot. Moreover, serving several purposes in a single excursion will raise the attractiveness of the centre as a destination.

The development of centres and the reduction of car use may be linked in this way. However, the development of a centre will also have its own negative aspects, including an increase in traffic congestion in surrounding areas as more and more cars are attracted. This is likely to worsen local air quality and have a negative impact on the reduction of CO_2 emissions and energy consumption, although some may stop using cars to avoid the congestion.[2]

1.3 Aim of this chapter

To enable future retail developments to adopt similar principles to those outlined above, an appropriate planning and control system is necessary. To this end, in the next section I will attempt to clarify any possible problems or missing elements in the Japanese retail development control system, in terms of restraint of car use, by comparing it with two similar land use control cases – one in England (Cambridgeshire) and one in the United States (Oregon, Portland) – which were partly designed for such a policy objective. Although these could be seen as relatively advanced systems, I would point out that this chapter has been based on a paper I originally wrote in the summer of 2001 (Takami 2002), and so it is possible that some of the information included may now be out of date. My discussion of these two case studies will follow on from the points set out in the next section.

2. Addressing retail development control systems

2.1 Ensuring commitment to the objective of restraining car use

Reducing car use is an issue that involves a wide geographical area, as it is often linked to the reduction of CO_2 emissions. However, land use is usually controlled by local municipalities, none of which is particularly interested in reducing car use and so they have little motivation to implement land use control for that purpose. We will therefore first focus our attention on the mechanism needed to gain commitment from each municipality in the region to the policy objective of car use reduction and to ensure planning and implementation in line with that objective.

2.2 Region-wide coordination on the scale of development, the characteristics of the centre and the level of parking provision

A large-scale retail facility with a wide catchment area may have a substantial impact on the pattern of shopping travel behaviour. For example, any restriction by a major regional city on retail development in its suburbs might diminish the vitality of its own city centre by diverting shoppers elsewhere if surrounding smaller municipalities attract large-scale development projects through "fiscal zoning" in order to revitalize themselves or increase tax revenue. In this case, car use could be expected to increase substantially if the retail facilities thus developed are not easily accessible by non-car modes.

   In order to avoid such consequences, the location of large-scale developments and the functions of any peripheral centres should not be decided solely by the individual municipalities but considered and coordinated from a broader perspective. Therefore, attention will also be focused on two major determinants for the likely attractiveness of proposed developments and centres – namely, the scale of development and characteristics of the centre, together with the level of parking provisions, reflected in plans prepared by municipalities as a result of coordination.

2.3 Regulation method and criteria, and information required

Regulation on developments in general, not solely on retail developments, is based on municipalities' reviews of applications submitted by developers. Decisions are made on predetermined criteria including the use and floor area ratio defined for each site in a zoning plan, or on a case-by-case basis focusing on the conformity with municipal policies. Although it is impossible to determine which approach is better, in the former, rigorous zoning regulations are necessary to control developments in detail. In the latter, judgement criteria need to be clarified in advance in order to exclude arbitrary decisions. In some cases, applicants are also required to submit specific information, such as the result of an assessment.

In the following case studies, therefore, attention will also be focused on regulatory methods and criteria, as well as the specific information required in reviewing development proposals.

3. An English case study: Cambridgeshire County

3.1 Overview of Cambridgeshire County

Cambridgeshire is a county located about 100 km north of London, and comprises four districts in addition to the principal city of Cambridge (Figure 6.2). The City of Peterborough, which also belonged to Cambridgeshire in the past and is situated within its boundaries, became a one-tier unitary authority in 1998. According to the geographical classification of central government, the county of Cambridgeshire and the City of Peterborough are both included in the Eastern England region.

Cambridgeshire County has a rural character as a whole. Research and development functions, including colleges and high-tech industries, are concentrated on the City of Cambridge. Historic sites at the centre of the city also attract a large number of tourists. A green belt extends from the suburbs of Cambridge to the district of South Cambridgeshire. The county is regarded as a forerunner in transport planning, with a suc-

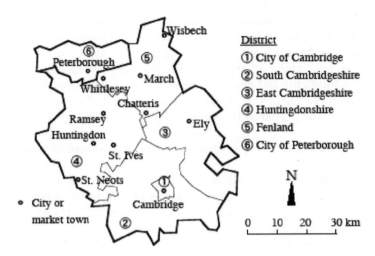

Figure 6.2 Cambridgeshire county (*Source*: created by the author)

cessful package approach integrating travel demand management and investments promoting public transport, pedestrian and cyclist facilities.

With a total area of some 3,055 km² and a population of about 0.53 million, the county is smaller than an average Japanese prefecture, both in terms of area and population. The "Cambridge Area", which refers to Cambridge and South Cambridgeshire combined, has a total area of about 940 km², with a population of some 0.24 million.

3.2 Planning structure

Before discussing the mechanism to define a policy objective of car use reduction and seek commitment at the local level, we need to understand the role of the central government, counties and districts in the system of planning (Table 6.1).

The central government makes national policies and publishes acts, strategies, guidances, etc. The Town and Country Planning Act provides for the planning system framework. The Environment Act requires the government to prepare the National Air Quality Strategy and provides for the mechanism of Local Air Quality Management. Laws closely related to the reduction of car use include the Road Traffic Reduction Act and the Road Traffic Reduction (National Targets) Act, which require the government and local authorities to set targets for reducing the level of road traffic (i.e. vehicle kilometres travelled). The central government has also set national targets on air quality, climate change (i.e. reduction of CO_2 emissions), modal share of bicycle, and so on.

Table 6.1 Major statutes, guidances and plans related to land use and transport in Cambridgeshire

Central government	Town and Country Planning Act
	The Environment Act, National Air Quality Strategy
	Road Traffic Reduction Act, Road Traffic Reduction (National Targets) Act
	Transport Act
	Planning Policy Guidance: PPG6 (Town Centres and Retail Developments), PPG13 (Transport)
	National targets: air quality, road traffic reduction, climate change (CO_2 emission), number of bicycle trips, etc.
(Sub-) Region	Regional Planning Guidance: RPG6 (RPG for East Anglia)
County	Structure Plan: Cambridgeshire Structure Plan 1995
	Local Transport Plan: Cambridgeshire Transport Plan
District	Local Plan
	Local Transport Plan: Peterborough Local Transport Plan

Source: Created by the author.

Furthermore, the central government issues Planning Policy Guidances (PPGs) to provide local authorities with non-statutory guidances for planning-related policies. Guidances on retail developments are provided in "PPG6: Town Centres and Retail Developments" (DoE 1996), while guidances on transport are provided in "PPG13: Transport" (DETR 2001).

At the regional level, Regional Planning Guidance (RPG) provides a comprehensive strategic policy for the next 15 to 20 years. Cambridgeshire is currently covered by RPG6 (DETR 2000). RPG6 covers the East Anglian region, which includes Cambridgeshire as well as the neighbouring counties of Norfolk and Suffolk.

The structure plan (SP) serves as a framework at the county level. The latest structure plan for Cambridgeshire County (Cambridgeshire County Council 1995) covers the area of the former county, including the City of Peterborough. At the district level, local plans (LPs), prepared by individual districts, provide detailed criteria for authorizing developments. SP and LP, as well as unitary development plan prepared by unitary authority, are collectively known as development plans (DPs).

Cambridgeshire County and the City of Peterborough also prepare separately a five-year local transport plan (LTP) concerning transport strategy and investments, which is submitted to the government in order to obtain funding for local transport schemes. Each LTP has to include performance indicators and targets to measure achievements, as well as relevant objectives, strategies and implementation programmes. Central government's decision on funding allocation is based on the contents of the LTP and the extent to which its objectives and targets have been achieved (Ohta 2000).

3.3 Ensuring commitment to the objective of restraining car use

The central government includes in its policy objectives the reduction of car use. PPGs present planning policies and measures to be taken for its achievement.

In summary, the objectives of PPGs regarding retail developments and transport include the revitalization of existing centres, the improvement of access to shopping centres and other facilities by a choice of means of transport, and the promotion of sustainable (i.e. non-automobile) transport choices. PPG6 also aims at maintaining an efficient, competitive and innovative retail sector, stating that it is not the role of the planning system to restrict competition, to preserve existing commercial interests or to prevent innovation. It should also be noted that PPGs do not provide any supply-side objectives, such as the development of transport infrastructure or the objective of ensuring smooth traffic flow, partly because PPGs are mainly guidances for land use planning.

As locational and transport policy measures to achieve those objectives, PPGs call for:
1. The location of major generators of travel in existing centres and at major public transport nodes;
2. The location of convenience shopping and other everyday opportunities at local centres, thus ensuring easy access for pedestrians and cyclists;
3. The encouragement of mixed use developments and the diversity of use at centres;
4. The supply of quality parking spaces for the revitalization of centres, although reducing the amount of parking is essential as part of a package of measures.

Local authorities have to follow PPGs and other central government policies at both stages of planning and implementation. Compliance with this requirement is ensured by the following mechanism. RPG is drafted by a Regional Planning Body (RPB), composed of local authorities in the region, in collaboration with regional offices and other relevant organs of the central government. The final RPG is issued by the central government after review and approval by the Secretary of State in charge. As for DPs, consultation starts at the initial stage of preparation with neighbouring local authorities and administrative organizations at both higher and lower levels. Although the approval for a prepared DP by the Secretary of State is not required, he has the power to recommend or order modifications of the DP if it is not consistent with national or regional policies. He has also the power to call-in a decision on approval of a planning application if the decision does not conform to the government policies, or if the proposed development is expected to be of more than local importance. Through this mechanism, the central government seeks to ensure harmonization of PPGs, RPGs and DPs, as well as the conformity of planning decisions with national policies.

In addition, local authorities are required to reflect in their LTPs, as closely as possible, the established national targets, such as air quality, reduction in CO_2 emission and the number of bicycle trips, when setting performance indicators and targets. When deciding on funding allocation, the central government considers the consistency of each LTP with DPs, PPGs and other national policies and objectives, as well as the extent to which its objectives and targets have been achieved. Thus, local authorities are also encouraged financially to give due consideration to the compatibility between the land use plan and transport plan, as well as between those plans and national policies, in preparing LTPs and DPs.

3.4 Region-wide coordination on the scale of development and centre characteristics

PPG6 requires RPGs to assess the scope for developing regional shopping centres (i.e. retail stores with a gross floor space of 50,000 m² or more). It requires SPs to set out the hierarchy of the centres, as well as the strategy for the location of employment, shopping, leisure and other uses that generate many trips. Identification of development sites is required at the LP level. This subsection outlines how these requirements are reflected in the prepared plans.

RPG6 makes it clear that there is no scope for an additional regional shopping centre during the period covered by the RPG because the growth of population, and hence retail demand, are slowing and spread across East Anglia.

The SP for Cambridgeshire defines two city centres and 12 town centres. The strategy adopted by the SP on the location of retail facilities provides that major convenience shopping developments (with a retail floor space of 1,400 m² or larger) should be located within or adjacent a centre defined above, or in a place within the City of Cambridge that is easily accessible without a car. It states that large-scale out-of-town shopping centres (10,000 m² or larger) will not be permitted outside the Cambridge Green Belt – that is, in most parts of the county excluding the City of Cambridge – as they have a substantial impact on smaller centres and the surrounding environment, and are difficult to access without a vehicle. As for the Cambridge area, the SP states that one or two large-scale shopping developments with a total retail floor space of 27,900 m^2 may be developed to alleviate congestion and meet future shopping demand. These developments are required to meet five requirements: high accessibility to the entire sub-regional catchment; good access to primary road network; ease of access by public transport, cycles and pedestrians; minimization of adverse effects on the environment; and provision of a high-quality retail facility. The SP proposes the Cambridge northern fringe area as a suitable development site.

The Huntingdonshire LP (Huntingdonshire District Council 1995) states that new shopping developments may be generally permitted that are unlikely to have an adverse effect on the vitality and viability of existing centres or on traffic movements. The LP requires individual shopping proposals to be satisfactory in terms of site selection, design, car parking, accessibility by various means of transport and environmental impact. In line with PPG6 and the Cambridgeshire SP, the LP specifically provides that a major convenience shopping development (with a retail floor space of 1,400 m^2 or larger) should be located within, or immediately adjacent to, the established town centre shopping area. Edge-of-town or out-of-town locations may only be permitted if suitable sites are not available on the said location. The LP also identifies one site for a superstore (2,500 m^2 or larger) and three sites for small shopping centres (smaller than 1,400 m^2) serving new residential neighbourhoods. Some of the sites are shown in the proposal map. Development of local shopping facilities to serve existing residential areas may be allowed if they do not conflict with other policies set out in the LP.

Thus, in England, RPGs and SPs are supposed to deal with relatively large-scale retail developments. It is the role of LPs to deal with smaller developments and to identify possible development sites. It is intended, through the system outlined above, to maintain the vitality of existing centres and to strike a balance between various types of centre.

3.5 Region-wide coordination on the level of parking provision

PPG6 requires that RPGs (if possible), or SPs, set the parking standards applicable to the whole region. PPG13 requires that RPGs include not only parking standards but also a consistent approach on parking, including policies related to parking charges and controls. Although the previous PPG13, issued in 1994, referred to both minimum and maximum standards, the present PPG13 changed this practice and stated that "there should be no minimum standards for development, other than parking for disabled people". Thus, the central government determines maximum parking standards at the national level (Table 6.2). Parking

Table 6.2 National maximum parking standards in England, according to PPG13

Use	National maximum parking standard (m^2 = gross floor space)	Threshold from and above which standard applies (gross floor space)
Food retail	1 space per 14 m^2	1,000 m^2
Non-food retail	1 space per 20 m^2	1,000 m^2
Cinema and conference facilities	1 space per 5 seats	1,000 m^2
D2 (other than cinemas, conference facilities and stadia)	1 space per 22 m^2	1,000 m^2
B1 including offices	1 space per 30 m^2	2,500 m^2
Higher and further education	1 space per 2 staff + 1 space per 15 students	2,500 m^2
Stadia	1 space per 15 seats	1,500 seats

Source: Department of the Environment, Transport and the Regions (2001).

standards have to be defined by use in DPs (particularly in LPs). Local authorities and RPBs may set stricter standards than the national standards.

At the time of writing, however, revisions to RPG6 and DPs in the county have as yet not been completed as the PPG13 itself has just been revised. To date, RPG6 has not yet defined maximum parking standards for the region, and the national standards will be applied for the time being.

3.6 Regulation method and criteria

In England, a prospective developer submits an application (including change-of-use application) to the Local Planning Authority, which decides on the permission of the application on a case-by-case basis (Planning Permission system). The decision is based primarily on relevant DPs (LPs in particular). Other related matters are also to be considered before a decision is reached.

PPG6 shows the list of matters to be considered in the planning permission process (Table 6.3). With regard to a development proposal for a large-scale store or a development proposal outside of the existing centres, key considerations include: accessibility by public transport, on foot and by bicycle; impact on travel and car use; and the possibility of adverse impacts on existing centres. This provision has a considerable effect on planning permission. Indeed, some development proposals have been rejected on the grounds that they would undermine the vitality of existing centres, that they did not adopt a sequential approach[3] (i.e. an-

Table 6.3 Matters to be considered in reviewing development proposals

Development proposal of major retail facilities (gross floor space exceeding 2,500 m^2)	
Sequential approach	Whether the applicant has adopted a sequential approach to site selection. Availability of suitable alternative sites.
Impact on centres	Likely economic impact on town centres, local centres and villages, including cumulative effects of recently completed developments and outstanding planning permissions.
Accessibility	Accessibility by a choice of means of transport. Proportion of customers likely to arrive by different means.
Travel pattern	Likely changes in travel pattern over catchment area.
Environmental impact	Any significant environmental impact (where appropriate).

Development proposal outside existing centres	
Impact on DP strategy	Whether the proposal would undermine the DP strategy.
Impact on the vitality and viability of existing centres	Extent to which the proposal would put at risk town centre strategy; likely effect on future private sector investment; changes to the quality, attractiveness and character of centres; changes to the physical condition of the centre; changes to the range of services provided by existing centres; and likely increases in vacant properties in the primary retail area. Likely effects on nearby centres if the proposed investment is not made.
Accessibility	Likely accessibility by a choice of means of transport; whether the proposal is genuinely accessible by non-automobile modes (not only the existence of public transport service but also the level of service). Prospective developer may be required to contribute to improving the public transport service and/or pedestrian access through planning obligations.
Impact on travel and car use	Whether the development would ensure easier access to all customers; facilitate more linked (multi-purpose) trips; and help achieve the overall aim of reducing reliance on the car for all trips. Likely proportion of customers who would arrive by car; the catchment area which the development seeks to serve.

Source: Department of the Environment (1996).

other suitable site exists within or nearer to a centre), or that the proposed site is not easily accessible by non-car modes. Accordingly, the Huntingdonshire LP sets criteria for permitting new retail development proposals, as described in the previous subsection.

As permission is made on a case-by-case basis, it is able to impose detailed regulations on use of developments. A proposal to develop retail stores in a suburban area, for example, may be restricted by the categories of goods to be sold by the stores through planning condition, so that any adverse effect on existing centres may be avoided. Such regulations are allowed by PPG6.

Even if the Local Planning Authority should give a permission to a retail development application that runs counter to PPG6, the Secretary of State may call-in and make a judgement if it is considered that the decision does not conform to relevant central government policies. PPG6 requires the Local Planning Authorities to report to the Secretary of State any retail development proposal with a gross retail floor space of 10,000 m^2 or larger. This requirement is intended to give him an opportunity to consider the possibility of call-in.

3.7 Information required

A developer proposing a development that may have substantial impact on the environment and/or traffic is required to submit Environmental Assessment, Transport Assessment and Travel Plan at the time of application (Table 6.4). These documents are taken into account in the planning permission process.

Environmental Assessment estimates the impact of a proposed development on the environment. Transport Assessment was introduced to replace Traffic Impact Assessment. While the latter limited itself to assessing how the concentration of traffic to the site will affect the surrounding road network, the former assesses the impact of the development on people's movement by various modes of transport. Environmental Assessment and Transport Assessment also include measures to be taken by the developer to alleviate the impact. Although Travel Plan primarily outlines measures to be taken by entities such as firms, schools and hospitals to reduce vehicle use, including commuting by single-occupant vehicles, it may also be applied to retail development proposals.

Depending on the result of Transport Assessment, planning permission may be given with a planning condition or obligation imposed, or after reaching a planning agreement, to require on-site transport measures or developer contributions, for instance. Developer contributions may be used for infrastructure development and service improvement of public transport, as well as for road construction.

Table 6.4 Additional information for reviewing development proposals.

Environmental Assessment	Required for development proposals with potentially significant impact on environment (e.g. out-of-town retail developments with a gross floor space exceeding 20,000 m^2).
	To be submitted at the time of application, describing: the development proposal outline; data needed for assessment; direct and indirect impact on environment; and measures to mitigate adverse effects.
	The term "environment" refers to human beings, animals, plants, soil, water, air, climate, landscape and the interaction of those elements, as well as material assets and cultural heritage.
Transport Assessment	Required for development proposals which will have significant transport implications (e.g. retail developments with a gross floor space exceeding 1,000 m^2).
	To be submitted at the time of application, illustrating: accessibility to the site by all modes; likely modal split of journeys to and from the site; and detail of proposed measures to improve access by non-automobile modes, reduce the need for parking associated with the proposal and mitigate transport impacts.
Travel Plan	Required for development proposals which are likely to have significant transport implications (e.g. retail developments with a gross floor space exceeding 1,000 m^2).
	A package of practical measures to be submitted at the time of application, aimed at the delivery of sustainable transport objectives, such as: reduction in car use (particularly single occupant journeys) and increased use of non-car modes; reducing traffic speeds for safety and security; and more environmentally friendly delivery and freight movements.
	The implementation of Travel Plan may be enforced by planning conditions or obligations.

Source: Department of the Environment (1996).

4. Oregon: Portland metropolitan area case study

4.1 Overview of Portland metropolitan area

Portland is the largest city in Oregon, located in the northernmost area of the state and on the opposite bank of Columbia River to the State

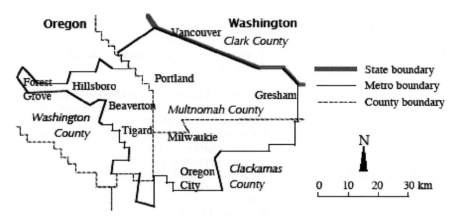

Figure 6.3 Portland metropolitan area (*Source*: created by the author)

of Washington (Figure 6.3). A city known as one of "the most liveable cities", Portland has succeeded in revitalizing its city centre since the 1970s by elaborating and implementing a Downtown Plan, focusing on the development of light rail and bus services, construction of transit malls, mixed-use developments on a human scale, and restriction on car parking in the city centre. Since the 1990s these attempts have been expanded to cover the whole metropolitan area, and a policy has been adopted to accommodate future growth along public transportation corridors in anticipation of rapid population growth. The area is said to provide a useful model of growth management.

A striking characteristic of Portland metropolitan area is the existence of a directly-elected regional government called "Metro" between the levels of state government and local governments (cities and counties). Comprising 24 cities in three northern Oregon counties, and with an area of some 1,200 km^2 under its jurisdiction, Metro's primary responsibility is to prepare regional plans and policies. About 1.3 million people live in the Metro area.

4.2 Planning structure

Table 6.5 lists the major laws, guidelines and plans developed by the federal, state and local governments, as well as Metro.

The federal government has issued major legislation including: the Clean Air Act Amendments (1990), the Intermodal Surface Transportation Efficiency Act (ISTEA) (1991) and the Transportation Equity Act for the 21st Century, which superseded the ISTEA in 1998. It has also

Table 6.5 Major laws, guidelines and plans related to land use and transportation in Portland metropolitan area, Oregon

Federal government	Transportation Equity Act for the 21st Century
	Clean Air Act Amendments
	National targets: air quality, etc.
State government	Statewide Planning Goals and Guidelines
	Oregon Administrative Rules, Oregon Transportation Planning Rule
	Oregon Transportation System Plan (State TSP), Statewide Transportation Improvement Program
Regional government: Metro	Regional Urban Growth Goals and Objectives
	2040 Growth Concept
	Urban Growth Management Functional Plan
	Regional Transportation Plan (Metro TSP), Metropolitan Transportation Improvement Program
	Regional Framework Plan
Local governments	Comprehensive plan
	Zoning ordinance
	Transportation System Plan (local TSP)

Source: City of Portland (1996).

provided financial assistance for the development of transportation infra- structure (Ohta 2000). In principle, federal government intervention in transportation and retail location policies at state, regional and local gov- ernment levels is limited.

The government of Oregon, under the Land Conservation and Devel- opment Act, requires all counties and cities to prepare a comprehen- sive plan. The state government has also introduced the Urban Growth Boundary (UGB) system to prevent urban sprawl. UGB comprises the land area for accommodating the growth expected over the next 20 years, and any developments outside the boundary are prohibited as a rule. The Land Conservation and Development Commission (LCDC) of the state government has prepared 19 Statewide Planning Goals and Guidelines, as well as the Oregon Administrative Rules (OARs) to implement them.

The government of Oregon has also prepared the State Transportation System Plan (TSP), which provides a vision of transportation in the state for the next 20 years. At the same time, the state government requires Metropolitan Planning Organizations, including Metro, as well as cities and counties to prepare a TSP in accordance with the Statewide Goals and OARs. TSPs prepared at each level are required to be consistent with each other and with relevant land use plans (comprehensive plans, for example).

In accordance with the Statewide Goals and OARs, Metro has developed the Regional Urban Growth Goals and Objectives, the 2040 Growth Concept, which provides a future blueprint of the region, the Urban Growth Management Functional Plan (Metro 1996) to put the Growth Concept in practice, as well as the Regional Framework Plan (Metro 1997), which synthesizes the above-mentioned and other relevant plans and policies.

Local governments have developed a comprehensive plan in line with the policies and plans of the state government, and Metro Zoning and other regulations have been prepared and implemented in accordance with the comprehensive plan. The state government requires local governments to ensure that the zoning code is consistent with the comprehensive plan, and that those comply with the Statewide Planning Goals and Guidelines.

4.3 Ensuring commitment to the objective of restraining car use

As described above, the LCDC has developed 19 Statewide Planning Goals and Guidelines in Oregon. The comprehensive plans developed by local governments are reviewed by the LCDC in order to ensure compliance with the Statewide Goals. The plans take effect only after compliance has been confirmed. When a plan is rejected, the state government may cut back on subsidies to the local government.

The Statewide Goals do not include any description on retail developments. Although automobiles are not singled out, the Goals include avoiding principal reliance upon any one mode of transportation, minimizing adverse social, economic and environmental effects, and conserving energy.

The purpose, stated in the division of Transportation Planning of OARs (Oregon LCDC 2001), which is also called the Transportation Planning Rule (TPR), is to promote the development of transportation systems that are designed to reduce reliance on the automobile so that air pollution, traffic and other liveability problems might be avoided. Thus, TPR requires Metro and local governments to design Metro TSP and local TSPs so that vehicle miles travelled (VMT) per capita may be reduced by 10 per cent within 20 years, and by 15 per cent within 30 years. As a means of reducing reliance on automobile and achieving the target, TPR requires local governments to allow Transit Oriented Development (TOD) ensuring coordination between land use and transportation, and to implement parking plans to achieve a 10 per cent reduction in the number of parking spaces per capita within 20 years. In terms of retail developments, the prohibition or restriction of scattered large-scale developments are not called for. It does state, however, that the local governments should, in the TSP process, evaluate and consider alternative land use plans which increase allowed densities in new office and retail developments in centres, and designate lands for neighbourhood shopping centres within convenient walking and cycling distance of residential areas in order to encourage access on foot or by bicycle.

Thus, integrated planning of land use, including retail developments, and transportation for curtailing car use (i.e. VMT) is not only a policy objective at the state level but also an obligation for local governments.

4.4 Region-wide coordination on the scale of development and the characteristics of centres

At the state level, specific policies on retail developments do not appear in the Statewide Goals nor TPR, except for increasing the density of centres and developing neighbourhood centres, as mentioned above.

In the 2040 Growth Concept, Metro gave shape to this guideline by defining "design types" including three types of centres (central city, regional centres and town centres) as well as corridors, main streets and station communities, which assume the functions of lower-level centres. Those centres have been identified on the Growth Concept Map (see Takami 1998 or 2000 for details). However, as the Growth Concept is only a concept, it has no binding force on plans prepared by local governments. The role to ensure implementation of the Growth Concept is assumed by the Urban Growth Management Functional Plan. The Plan provides the characteristics and the recommended density of each design type as well as various standards and policies, including regional parking standards, and requires local governments to amend their comprehensive plans so that each design type may be applied in line with the Growth Concept Map. In Employment Areas and Industrial Areas, retail uses are limited to the level necessary to serve the needs of businesses and employees. In principle, the Plan requires that comprehensive plans and implementing regulations be amended to prohibit retail uses larger than 60,000 square feet (approx. 5,600 m^2) of gross leasable area.

Although Metro restrains development outside the UGB, it is unlikely that this intends especially to prevent suburban retail developments, as the UGB is defined for the purpose of proper accommodation of the growing population.

The policies and plans described above are also considered in comprehensive plans prepared by local governments. Although the comprehensive plan of Portland (City of Portland 1996) includes little reference to policies related to retail developments, five types of commercial zones appear on the plan map: neighbourhood commercial, office commercial, urban commercial, general commercial and central commercial. The map is roughly consistent with the 2040 Growth Concept, as many of the commercial zones are located in centres and corridors identified in the Growth Concept. The zoning map for land use regulations is also based on this comprehensive plan map.

4.5 Region-wide coordination on the level of parking provision

At the state level, TPR sets a goal of achieving a 10 per cent reduction in the number of parking spaces per capita within 20 years, and requires local governments to include land use regulations setting minimum and maximum parking requirements in appropriate locations, and to reduce the minimum requirement. No statewide standards have been established.

Metro's Urban Growth Management Functional Plan provides for parking minimums and maximums by development type (Table 6.6). It classifies the Metro area into Zones A and B according to the accessibility by public transportation or walking: parking maximums are lower in the more accessible Zone A, and higher in the less accessible Zone B. Local governments may set lower parking minimums and maximums to those stipulated in the Plan.

In the City of Portland, the zoning code provides for minimum and maximum parking standards by zone and by use. Thus, the number of parking spaces are controlled within the framework of zoning regulations. It also increases the maximum standards in less accessible areas by public transportation, and reduces the minimum standards for developments incorporating the development of transit-supporting plazas.

4.6 Regulation method and criteria

In Oregon, as in other states in the US, the state government enables local governments to implement zoning regulations by ordinances, providing a positive list of buildings that may be constructed. It is said that this allows meticulous regulations that meet the specific conditions of each area, as numerous zoning categories have been established. However, zoning changes more frequently than in Japan. The zoning of a site may be changed to accommodate a development after due process, including public hearings.

Zoning is defined in line with a comprehensive plan. The zoning code of Portland (City of Portland 2001a) defines 27 base zones including six single-dwelling residential, six multiple-dwelling residential (both classified according to minimum lot size, etc.), eight commercial, three employment, three industrial and one open space. In addition, 18 overlay zones may be established on base zones. Allowed uses, floor area ratio, height, setbacks, building coverage and parking standards are prescribed for

Table 6.6 Regional parking standards for Portland metropolitan area, according to Metro's Urban Growth Management Functional Plan (excerpts)

Land use	Minimum required	Maximum permitted: Zone A	Maximum permitted: Zone B
General Offices (gross floor area)	2.7	3.4	4.1
Light Industrial, Industrial Parks, Manufacturing (gross floor area)	1.6	–	–
Warehouses (gross floor area, over 150,000 square feet only)	0.3	0.4	0.5
Schools: College/University and High Schools (spaces/number of students and staff)	0.2	0.3	0.3
Sport Clubs, Recreation Facilities	4.3	5.4	6.5
Retail/Commercial, including shopping centers	4.1	5.1	6.2
Movie Theaters (spaces/number of seats)	0.3	0.4	0.5
Medical/Dental Clinic	3.9	4.9	5.9
Hotel/Motel	1	–	–
Single Family Detached	1	–	–
Multi-family, townhouse, one bedroom	1.25	–	–
Multi-family, townhouse, two bedroom	1.5	–	–
Multi-family, townhouse, three bedroom	1.75	–	–

Sources: City of Portland (1996); and various reports.
Notes:
Unless specified otherwise, figures indicate the number of parking spaces for a gross leasable area of 1,000 square feet.
Minimum required parking standards for downtown Portland are provided elsewhere.
Zone A includes areas easily accessible by public transportation or for pedestrians. Zone B includes other areas.

each of the commercial base zones (Table 6.7), which correspond roughly to the five zones defined in the comprehensive plan. Since "retail sales and service" represents a single land use, it is unable to apply different regulations on the location of stores depending on, for example, the items they deal in.

However, the fact that the regulations are based on zoning does not necessarily mean that they are completely rigid. There is still room for discretion. The zoning code of Portland provides for allowed, limited, conditional and prohibited uses for individual zones. Conditional uses

Table 6.7 Characteristics of commercial zones (City of Portland)

Zones	Characteristics
Neighbourhood Commercial 1	Small retail sites in or near dense residential neighbourhoods. Pedestrian-oriented. Parking areas are restricted.
Neighbourhood Commercial 2	Small commercial sites and areas in or near less dense residential neighbourhoods. Predominantly auto-accommodating except where the site is adjacent to a transit street.
Office Commercial 1	Low intensity office zone that allows for small-scale offices in or adjacent to residential neighbourhoods.
Office Commercial 2	Low and medium intensity office zone with a local or regional emphasis. Somewhat auto-accommodating except where the site is adjacent to a transit street.
Mixed Commercial/ Residential	Development intended to consist primarily of businesses (locally oriented retail, service, and office uses) on the ground floor with housing on upper stories. Located on busier streets and pedestrian-oriented.
Storefront Commercial	A full range of retail, service and business uses with a local and regional market area intended to preserve and enhance older commercial areas that have a storefront character. Industrial uses are allowed but are limited in size. Pedestrian-oriented.
General Commercial	A full range of retail and service businesses with a local or regional market. Industrial uses are allowed but are limited in size. Generally auto-accommodating except where the site is adjacent to a transit street.
Central Commercial	Very intense commercial development within Portland's most urban and intense areas. A broad range of uses is allowed. Pedestrian-oriented with a strong emphasis on a safe and attractive streetscape.

Source: City of Portland (2001a).

are allowed provided that they are approved in a special review process that examines whether they comply with a set of criteria and other regulations. Approval criteria differ for individual uses and zones, including, for example, that the transportation system is capable of supporting the proposed use and the existing uses in the area in terms of street capacity and level of service, on-street parking impacts, pedestrian safety and so on; that the proposal will not have significant adverse impacts on the uses of nearby areas; or that the proposed use will not significantly alter the character of the area. The same proposed use in the same zone may be classified as allowed, limited or conditional according to the scale of development. For instance, small-scale retail uses are allowed as a conditional or limited use in high-density multi-dwelling zones.

In practice, many of the commercial zones are allocated in centres and along major public transportation (bus or light rail) lines. The overwhelming majority of the city area, excluding centres and open spaces, is designated as single-dwelling zones, where no retail stores may be located. Therefore, the opportunity is considerably limited in terms of locating a large-scale retail development in an area that is not well served by public transportation.

4.7 Information required

In Portland, a prospective developer who proposes a development of a certain scale (in terms of expected concentration of vehicle trips, etc.) may be required, at the initial stage of review, to submit the result of Transportation Impact Study, in which the impacts of the development on surrounding roads are assessed (City of Portland 1997). It focuses on estimating the number of vehicular trips attracted to the proposed development as well as ensuring smooth traffic flow on surrounding roads and safe pedestrian and cyclist access to the site. Unlike the Transport Assessment in England, it does not specifically call for the promotion of multi-modal transportation.

As a type of impact fee, the City of Portland has established the System Development Charge (SDC) scheme that requires all developers, in principle, to share the cost of transportation infrastructure development and improvement according to the use and scale of the development (City of Portland 2001b). For retail developments, the SDC is calculated as the product of the square footage of the development and the unit rate, which is set by the types of store (e.g. shopping centres, supermarkets, convenience markets). The SDC scheme also has various exemption measures. A new development located in the central city or near bus or light rail lines may obtain up to 65 per cent or 35 per cent reduction in the SDC, respectively. The applicant may also be eligible for a credit against SDC obligation by participating in certain types of street improvements. These SDCs are primarily invested in the development of North–South light rail lines, updating bus scheduling and location systems, improving streets and intersections with cyclist or pedestrian lanes, and constructing overpasses. Thus, the transition to TOD-like urban patterns is not only enforced by regulatory measures but also encouraged through the SDC scheme.

5. What is missing in the Japanese system?

The Cambridgeshire and Oregon cases outlined above indicate that several things are missing in the Japanese retail development control system. It will be worth noting the following five missing points.

5.1 Missing point 1: policy objective of car use reduction set at higher government levels

The two case studies show that the reduction of car use is defined as a policy objective at higher government levels, by the central government in England, and the state government in Oregon. In both cases, there are, or will be, numerical targets on reducing vehicle mileage.

In Japan, the Basic Environment Plan, developed by the former Environment Agency and published by a Cabinet Meeting Decision in December 2000, presents in the section on "Efforts toward Environmentally-friendly Transportation" the objective of achieving environmental standards and the greenhouse gas reduction target (which is the same as the Kyoto Protocol target of reducing emissions by 6 per cent from the level of 1990). Although the plan states that "comprehensive measures including the adjustment and reduction of automobile traffic demand" shall be taken, the reduction of car use itself, in terms of total vehicle kilometres travelled for example, is not included in the objectives. As a result, there has, to date, been no reduction target set.

The setting of policy objectives should provide the foundation for all policy measures. Thus, it may be noted that no strong foundation exists for policy measures to reduce car use.

5.2 Missing point 2: recognition of land use control as a policy measure

In England, PPG13 and PPG6 by the central government provide guidances on how to control land use, including retail developments, in order to reduce automobile use. In Oregon, the state TPR requires coordination between land use and transportation, adopting a TOD-based urban pattern. In both cases, coercive and inductive mechanisms are employed in addition to consultations with local municipalities to ensure that they develop plans and policies that meet policy objectives. Such mechanisms include recommendations and orders to revise plans, planning approval and controls through funding.

In Japan, controlling land use was long considered as one of the travel demand management measures solely to ensure smooth road traffic. The above-mentioned Basic Environment Plan not only provides for traditional measures for improved performance of individual automobiles, smooth traffic flow and efficient freight transportation, but also includes for the first time measures to reduce traffic demand itself by promoting public transportation systems and achieving environmentally-friendly urban structures. In other words, land use control measures have come to be recognized by the government as a tool for addressing environmental issues deriving from automobile traffic, as well as ensuring smooth road traffic flow.

However, there have been no land use control measures included in the list of measures with targets, defined by the Ministry of Land, Infrastructure and Transport, to be taken in the transport sector in order to achieve a 6 per cent reduction in greenhouse gas emissions. Nor do the City Planning Operational Guidelines, developed by the ministry, provide any vision for car use reduction. Furthermore, the system based on the Large-scale Retail Stores Location Law, under the Ministry of Economy, Trade and Industry, has little to do with issues related to wider areas, being designed solely to conserve the living environment in local communities.

Thus, it is true that the Basic Environment Plan provides a step towards restraining car use for addressing global environmental issues and recognizes land use control as a measure for it, although that is not listed as an objective. Nonetheless, this attitude has not been sufficiently reflected in the national line of policy, particularly in policies under the jurisdiction of other ministries.

5.3 Missing point 3: mechanism of region-wide coordination

In England, PPG6 ensures that the locational policies on major retail developments with widespread impact are considered from a regional viewpoint and included in RPGs or SPs, while policies on small-scale developments are considered at the district level and reflected in LPs. In Oregon, the configuration of centre-oriented land uses with various characters (e.g. three-layer centres) is shown in Metro's 2040 Growth Concept and Urban Growth Management Functional Plan, which is faithfully reflected in the comprehensive plans and zoning regulations of local governments.

As regards parking, appropriate levels of parking spaces are considered at the regional level in light of relevant objectives such as the restraint of overall car use and the preservation of the vitality of existing centres. In both cases, maximum allowed levels are defined in region-wide plans, based on the basic policy of putting a curb on parking spaces.

Thus, it may be understood that issues deemed to be deeply related to the competition among municipalities for developments are considered from a regional point of view in both systems.

In Japan, the Large-scale Stores Location Law has a mechanism at the prefectural level that demands developers and retailers of large-scale stores (with a retail floor space 1,000 m^2 or more) to pay attention to the environment of surrounding areas. The law, however, prohibits account to be taken of regional supply-and-demand of retail floor space during implementation because it is thought to inhibit competition among retailers. This makes it almost impossible to ask developers or retailers for any revision to the scale or location of a project even if problems can be foreseen. Thus, it seems impossible to utilize this system to ensure region-wide coordination. In the final analysis, the undesirable location of development may only be discouraged under a city planning system. The City Planning Operational Guidelines authorize zoning of city planning areas in order to deal with the dispersed location of large-scale developments. It may be possible to explore other tools (for example, the City Planning Area Master Plan introduced by the City Planning Law 2000) for region-wide coordination.

Sufficient parking spaces are to be supplied by mandatory minimum parking requirements and the Large-scale Stores Location Law. They are both designed to ensure smooth road-traffic flow in surrounding areas and have little to do with issues on wider areas and the reduction of overall car use. Therefore, the definition of maximum parking standards is completely outside the scope of those systems. To make matters worse, requiring excessive parking spaces in central sites can even encourage the suburban location of developments and, consequently, more car use. Setting aside the level of existing parking provision which is considerably less than the US and European cities, Japan is in striking contrast with the two metropolitan areas in terms of policy direction.

5.4 Missing point 4: a land use regulation system with effective control over developments

In England, development applications are reviewed on a case-by-case basis, allowing flexible decisions on planning permission and its conditions following the detailed examination of the impact of development. The goods or services to be provided in the stores may be limited to avert any adverse effects on existing centres. As in Japan, the zoning system is adopted in Oregon, but more meticulous regulation is possible due to the larger number of zone categories. Conditions may be attached to developments of a certain scale with certain uses, thus leaving some room for discretionary judgement. In the city of Portland, economic incentive is provided to encourage TOD-like urban patterns by imposing different charges (SDCs) on developers according to the uses and location of the development.

In Japan, large-scale developments in loosely regulated or unregulated zones, including unzoned areas and areas outside city planning area, have posed major problems. The 2000 revision to the City Planning Law addressed this issue by creating a system of "specific uses restricted areas", where municipalities may restrict uses of developments, and "quasi-city planning areas". However, effective control over retail developments depends on appropriate implementation of the system by municipalities, as the zoning code defined by the government lets large-scale stores to locate pretty freely, even in residential or industrial areas.

5.5 Missing point 5: consideration of transportation-related issues in reviewing development applications

In England, PPG6 requires consideration of the impacts of retail developments on overall travel patterns and car use in the decision of planning permission, effectively precluding developments that would increase car use. In Oregon, the impact on overall car use is not included in the assessment criteria, although the system of zoning regulation leaves some room for discretionary judgement, as was mentioned above. In both cases, the submission of assessment results on environment and transport and a plan for mitigating the impact of the development may be required in order to provide the necessary information for making a decision.

In Japan, building inspectors are required to permit applications which conform to prescribed regulations, such as use and floor area ratio, and are prohibited from making discretionary judgements in respect of other factors, including the likely effects on transportation.

In addition, developers or retailers of large-scale stores are required to conduct a variant of Traffic Impact Assessments under the Large-scale Stores Location Law. Some prefectures and cabinet-order designated cities have also issued ordinances that require developers to conduct environmental impact assessments. As described above, the former mechanism is focused on securing sufficient parking to prevent congestion on surrounding roads, and is not designed to reduce overall car use. Even if substantial impacts are foreseen, the two mechanisms cited above cannot disallow the development as they are outside the framework of city planning and building permission.

6. Conclusion

Suburbanization of retail developments can encourage automobile use, make a car-less life inconvenient and difficult, and consequently undermine some aspects of sustainability. In order to prevent this happening, the proper control of developments is necessary. From such a perspective, this chapter has examined retail development control systems in England and the United States and illustrated five missing points in Japanese system.

Problems derived from increasing car use should be tackled with comprehensive efforts, including demand management as well as supply-side measures. Land use control is classified as a long-term demand-side measure and should be a crucial element of a package of measures. In Japan, however, this kind of measure has not been effectively introduced to date. I think the reasons for this include: the defects in a system that hinders effective development control; the inadequate recognition of the effectiveness of land use control measures, partly due to insufficient research in this area; and, more fundamentally, the general lack of awareness of the need to move away from over-reliance on the car and the introduction of measure to restrain its use. We researchers should deal with these matters through further researches and enlightenment.

Notes

1. Efforts for strengthening the urban centre and preventing further retail developments in the suburbs should primarily aim at preserving the centre as a distinct area with particular functions, rather than ensuring the survival of specific retailers. However, there are strong grounds for the criticism that such a policy will benefit particular retailers located in the centre as well as those already established in the suburbs. Thus, measures need to be taken to ensure that locational policies will not restrict competition among retailers.
2. As I describe later in the chapter, in the Portland metropolitan area, one of the alternatives in the model analysis conducted in the process of preparing the 2040 Growth Concept was deemed most effective in controlling both vehicle miles travelled and air pollution. However, that alternative was also expected to result in substantial aggravation of traffic congestion. The finally recommended proposal incorporated the urban structure envisaged in the alternative, but called for the development of road infrastructure and partial absorption of population growth by satellite cities. Thus, traffic problems will be alleviated, although vehicle mileage will be somewhat greater (but smaller than the trend) (Takami 2000). In England, a Planning Policy Guidance (PPG13) provides that traffic management measures should be applied, such as reallocation of road space to pedestrians, cyclists and public transport and traffic calming, in order to avoid or manage congestion pressures in city centres caused by locational policies. These are inspiring examples in that they seek to reconcile land use control measures with anti-congestion measures, rather than giving up on land use control altogether due to the possibility of aggravating traffic congestion.
3. In selecting the site or potential sites of developments which are likely to attract many visitors, such as retail and leisure complexes, a sequential approach gives priority to (1) town centre sites, (2) edge-of-centre sites, and (3) out-of-centre sites that are in locations accessible by a choice of means of transport, in that order.

REFERENCES

Cambridge County Council (1995) Cambridgeshire Structure Plan, Cambridge, UK.

City of Portland (1996) Comprehensive Plan Goals and Policies, Portland, Oregon.

───── (1997) Transportation Impact Study Guidelines: Final Draft, Portland, Oregon.

───── (2001a) Portland Zoning Code, Portland, Oregon.

───── (2001b) Transportation System Development Charge, Portland, Oregon.

Department of the Environment (1996) PPG6: Town Centres and Retail Developments, London: HMSO.

Department of the Environment, Transport and the Regions (2000) Regional Planning Guidance Note 6: Regional Planning Guidance for East Anglia to 2016, London: Stationery Office.

───── (2001) PPG13: Transport, London: Stationery Office.

Huntingdonshire District Council (1995) Huntingdonshire Local Plan, Huntingdon.

Metro (1996) Urban Growth Management Functional Plan, Portland, Oregon.

───── (1997) Regional Framework Plan, Portland, Oregon.

Minohara, Kei, Yoshiki, Kawai and Tadahiko, Imaeda (2000) *We Can't Do Without Downtown: What is Central City Invigoration?* Gakugei Shuppan Sha (text in Japanese).

Oregon Department of Land Conservation and Development (2001) OAR Division 12: Transport Planning, Salem, Oregon.

Ohta, Katsutoshi (Project Manager) (2000) *Urban Transportation Planning System in Japan, USA and England: Towards a better integration with urban planning and environmental policies*, Japan Research Centre for Transport Policy (text in Japanese).

Takami, Kiyoshi (1998) "Land Use and Transport (1): Car Use Reduction through Land Use Control", *Urban Study Series 17: City and Habitation: Reflection on Land, Housing and Environment*, Tokyo Metropolitan University Press, pp. 203–236 (text in Japanese).

───── (2000) "A Case Study of Portland Metropolitan Area in Oregon, USA", *A Basic Study on Urban Transport Strategies for Environmentally Friendly City*, Japan Research Centre for Transport Policy (text in Japanese).

───── (2002) "A Study on Retail Development Control System for Reducing Car Use: Based on Cases of England and Oregon, USA", *Comprehensive Urban Studies* 77: 5–19 (text in Japanese).

───── Yasunori, Muromachi, Noboru, Harata and Katsutoshi, Ohta (1998) "Argument on Land Use/Transportation Measures for Reducing Car Use and a Study on an Issue of Retail Location", *Infrastructure Planning Review* 15: 217–226 (text in Japanese).

7

Population stability and urban area

Hidenori Tamagawa and Nobuhiro Ehara

1. Polysemy of sustainability and urban population

Sustainability has become the main focus of attention among the global environment issues. "The Limits to Growth" (Club of Rome 1972) was a landmark study on this theme. Thirty years after its publication, the report is still regarded as a pioneering investigation, presenting analyses and forecasts of global environmental issues using quantitative simulations. Recent global policy efforts, including the Earth Summit, seem to draw on its conclusion that the state of global equilibrium will only be possible if we combine technological responses, including pollution control and resource recycling on the one hand, and systematic control on growth, particularly the stabilization of population and capital, on the other.

Five years after the report was published, the Third National Comprehensive Development Plan (National Land Agency 1978) was adopted in Japan against the backdrop of slower economic growth following the oil crisis. Launching the "suitable area for habitation" initiative, the plan called for "comprehensive environmental considerations for human settlement", modelled on river basin communities. This idea completely changed the concept adopted in the First and Second National Comprehensive Development Plans, which focused on industrial development and large-scale construction projects. "Energy conservation" became a buzzword in Japan during this period.

As environmental economics emerged as a discipline in the 1980s, the term "sustainable development" was first introduced in the 1987 World Commission on Environment and Development (WCED) report "Our Common Future". The report, also known as the Bruntland Report, suggests that developed donor countries need to sustain a certain level of economic growth if we are to solve environmental problems in developing countries. Apparently, this argument reconsiders the issue of environmental and economic development on a somewhat positive note, as compared with "The Limits to Growth". Subsequently, the term "sustainability" became recognized all over the world.

Many local governments have come to recognize sustainability as a policy issue. For example, the Tokyo Metropolitan Government has developed a Draft Plan of Action for a Recycling-oriented Society (1998), which provides guidelines on major issues such as resource recycling, the water cycle, energy conservation, transportation demand management (TDM) and the promotion of environmental learning. The current National Comprehensive Development Plan, "The Grand Design for the 21st Century", is also very much aware of the need for symbiosis with nature. Indeed, building nature-rich residential areas is one of its main objectives.

Any discussion on such themes hinges on the definition of the key term, sustainability. The term usually refers to the ecological sustainability of both nature and human beings, obtained as a result of controls on human activities. There would be no objection to adopting the term "ecosystem" as a key concept when discussing urban sustainability. Pielou's analysis (1969) revealed the diversity and contiguous order of vegetation. The comprehensive works of Abe et al. (1997) described self-organizing processes of ecosystem. These studies show the fundamental instinct of ecosystem. Some direct analogies by urban planners such as Geddes, Doxiadis and Hardin are discussed in chapter 10.

As far as we know, however, the phrase "urban ecosystem" may be discussed from at least two viewpoints.

One of the viewpoints considers the city as being part of a natural ecosystem. This may be called the natural scientist's concept. The other regards the city itself as an ecosystem, as symbolized in the expression "urban ecology". Burgess et al. (1925) was a frontier work of this field, succeeded by social area analysis by Shevky and Bell (1955) and sophisticated as a procedure by factorial ecology which identifies social areas as well as finding potential factors. The work of Herbert and Johnston (1976) and Davies (1984) are typical examples. This may be called the social scientist's concept.

Both views have an important role to play in planning theory and a similarity to each other as Jane Jacobs (2000) pointed out. Any action would necessarily result in a policy that ensures a balance between the two "ecosystems". In that case, however, two types of solution can be envisaged: taking account of the latter within the limit of the former, or vice versa.

The term "sustainability", as employed in the Japanese media, generally refers to the natural scientist's concept, for it is easier to build consensus around policies based on this concept for sustaining global environment. However, the pursuit of this ideal tends to culminate in extreme objectives such as dispersing self-sufficient villages nationwide or conversely, building skyscrapers in the middle of vast natural landscapes. Even setting aside its impracticality, this attitude overlooks the inevitable and necessary process of the creation and transformation of an intensive city with its concentrated population.

For this reason, this chapter starts the discussion from the latter interpretation of the concept of sustainability.

Some might criticize this approach as undermining recent progress in nature–human system discussions by bringing back anthropocentric viewpoints. I would like to counter this criticism by noting that an area that is no longer "sustainable" for anthropogenic use will not simply regain its status of "sustainable" land as a natural environment, and that desertion of such an area will generally lead to new encroachments of the natural environment. People's activities in the city may be classified into categories such as employment, living, learning and leisure, but living definitely has the lion's share in terms of the area required. Thus, the stability of the whole city largely depends on whether its residential areas are stable or subject to repeated desertion and dispersion.

According to the concept put forward by Common (1995), "sustainability" in an ecological sense implies "resilience" rather than "stability". In other words, it means "the ability to sustain systemic structure and behavioural pattern against any disturbance", and does not necessarily guarantee population stability of any species.

As he notes, however, it is not easy to find an indicator of "resilience". Furthermore, even if an area continues to be used, any decline in its habitation function may result in the above-mentioned encroachment of the natural environment in other areas, either through sprawling or systematic construction of new towns. It can thus be argued that "stability" is an important concept for individual urban areas. Indeed, the model provided in the Club of Rome report, mentioned earlier, included in its goals population stability, albeit on a very different scale. Thus, this chapter tentatively interprets "population stability" as indicating the "sustainability" of an area, focusing on the function of "habitation".

Recently, realistic and practical research has been undertaken on human settlement, particularly in the United States (Kawamura and Kokado 1995), but to date, there has not been adequate verification of the data. To address this gap, this chapter reviews the concept of "sustain-ability", using data as its main tool.

In summary, this chapter regards population stability as the expression of sustainability by focusing on the function of habitation, and seeks to provide, through positive data analysis, a tentative assumption on how to approach sustainability in the urban context, particularly that of a metropolis.

2. Changing population distribution and stable population districts in the Tokyo area

Based on the perception presented in the previous section, this section examines population stability in the Tokyo area, in other words, for the purposes of this chapter, the Metropolis of Tokyo as well as Chiba, Sai-tama and Kanagawa prefectures, as well as the southern part of Ibaraki prefecture.

Population change in the Tokyo area can be analysed for the period from 1970 to 1995, using the Basic Grid Square Statistics compiled from the Population Census. (1970 was the year when the Grid Square Statis-tics were first introduced in a systematic manner.) Since it has been agreed that the concentration of population on metropolitan areas reached a turning-point in about 1970 (Oe 1995), this time period ap-pears to be appropriate for identifying the characteristics of population change in an urban society. Although 25 years is a rather short period, Japan experienced social and economic upheavals throughout this time frame, including rapid economic growth, recession in the wake of the oil crisis, stable growth, the bubble economy and post-bubble recession. The total Metropolis of Tokyo population dropped after peaking in 1985. Subsequently, the pattern of population decrease in the 23 wards and the population increase in the Tokyo area (the Metropolis of Tokyo plus three prefectures) during the bubble period gave way to an increase in the wards and a decrease in the Tokyo area during a few years of the post-bubble period.

The rate of population change (DP) over the 25 years is calculated as follows:

$$DP(\text{per cent}) = 100(P_{1995} - P_{1970})/P_{1970},$$

where P_t: population in year t.

Figure 7.1 shows the population growth rate calculated for each of the grid squares (approximately 1 km × 1 km) that has not experienced a zero population year.

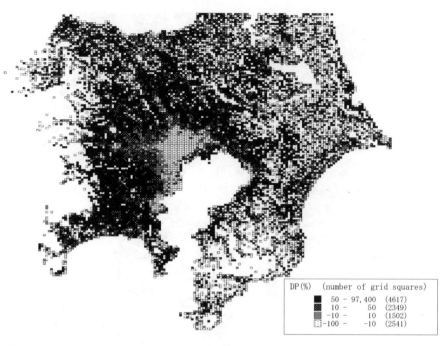

DP (%)		(number of grid squares)
50 –	97, 400	(4617)
10 –	50	(2349)
–10 –	10	(1502)
–100 –	–10	(2541)

Figure 7.1 Population change (DP: %) in Tokyo area over the last 25 years (*Source*: Produced by the author using data from Population Census by the Statistics Bureau, Ministry of Internal Affairs and Communications)

Grid squares with negative population growth are found at the centre of Tokyo, surrounded by districts with little population change. Grid squares with a gradual increase in population are located to the west of the city centre, which are in turn surrounded by districts with rapid population increases. Uneven population change can be observed in relatively remote districts in Chiba and Ibaraki prefectures. These were originally rural districts with small populations. Grid squares with little change or rapid growth in population can mostly be found to the west of Tokyo Metropolis. Such districts are relatively few to the north (Saitama) and to the south (Kanagawa) of Tokyo. Thus, the patterns of population change are not distributed in a concentric cycle.

As another indicator, the absolute sum of population change (absDPsum) between five-year periods (within the same 25-year period) is calculated as follows:

$$\text{AbsDPsum(per cent)} = 100 \cdot \Sigma |P_{t+5} - P_t|/P_t$$

where P_t: population in year t, t = 1970, 1975, 1980, 1985, 1990

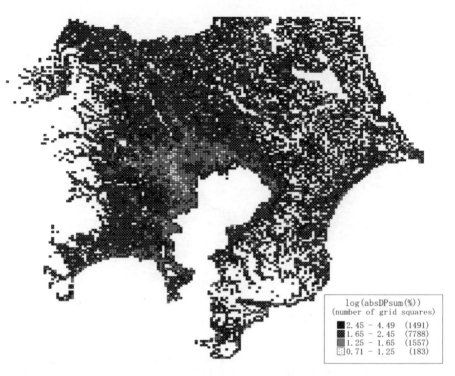

Figure 7.2 Logarithmic value of absolute sum of population change at 5-year intervals: log (absDPsum) (*Source*: Produced by the author using data from Population Census by the Statistics Bureau, Ministry of Internal Affairs and Communications)

Figure 7.2 classifies the grid squares according to the result. (classification based on logarithmic transformation to facilitate understanding).

Similar to the distribution found in Figure 7.1, grid squares with the smallest rates of change are chiefly found in the ward area to the west and southwest of the city centre. The boundary largely corresponds to the DID (densely inhabited district) boundary as of 1960, which indicates that those districts had been urbanized before motorization had become well established in Japan. Grid squares with minimum change in population can also be found along the JR (Japan Railway Company) Chuo Line as well as along the coastline extending from Kawasaki to Yokohama. Those districts are surrounded by grid squares with moderate population change, which are found in other parts of Tokyo Metropolis as well as in the coastal area of Kanagawa prefecture, including Kawasaki and Yokohama. The other grid squares may be defined as districts that experienced substantial changes in population (see Figures 7.3 and 7.4

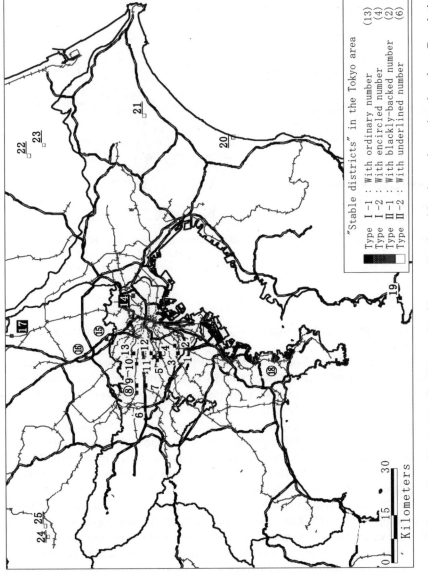

Figure 7.3 "Stable districts" in the Tokyo area (*Source*: Produced by the author using data from Population Census by the Statistics Bureau, Ministry of Internal Affairs and Communications)

173

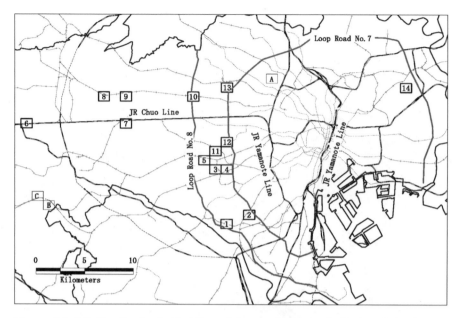

Figure 7.4 "Stable districts" (1–14) and "control districts" (A, B, C) in Tokyo Metropolis (*Source*: Produced by author using data from Population Census by the Statistics Bureau, Ministry of Internal Affairs and Communications)

and "Geographical background" of this book's beginning for the specific name of places).

As a measure of stability, let us abstract those areas where the absolute value of population change over the 25 years and the absolute sum of population change between five-year periods do not exceed 10 per cent. That is to say, little change can be observed in those districts in terms of difference between the beginning and the end of the 25-year period as well as the sum of the five quinquennial changes in population. The number of grid squares that meet these criteria is quite small: only 25 as shown in Table 7.1 and Figure 7.3. For the purpose of this chapter, those grid squares are tentatively defined as "stable districts".

Of the 25 stable districts, seventeen districts have a population density in the range of 10,000/km^2, while population density does not reach 1,000/km^2 in six of the districts. Two districts have experienced population increases in each of the five quinquennial periods, whereas one district's population dropped consistently. Many of the districts are to the west of the city centre. Others are dispersed in areas located more than 50 km from the city centre. Eighteen of the 25 districts have a railway

Table 7.1 Population trends in stable districts

No.	1970	1975	1980	1985	1990	1995	absDPsum	DP	75DID	Station
1	16215	16207	15787	15503	14913	14655	9.98	−9.6	1	Oyamadai
2	17980	17954	17774	18214	17495	17131	9.65	−4.7	1	Toritsu-Daigaku
3	17050	17792	17757	17902	17742	17089	9.94	0.2	1	Matsubara
4	17455	16728	16391	16593	16666	16369	9.63	−6.2	1	Umegaoka
5	16480	16121	15824	15637	15425	15636	7.93	−5.1	1	Kamikitazawa
6	13140	12912	12962	13046	12575	12904	9.00	−1.8	1	Nishi-Kunitachi
7	13800	14133	14346	14643	14787	14808	7.12	7.3	1	Musashi-Sakai
8	11100	10864	11212	11150	11156	10886	8.36	−1.9	1	Hanakoganei
9	12415	12294	12736	12561	12512	12443	6.89	0.2	1	Tanashi
10	12995	13353	13594	13752	13310	13276	9.19	2.2	1	Iogi
11	13515	13337	12889	12581	12630	12352	9.66	−8.6	1	Eifukucho
12	17485	17956	18218	18712	18997	19114	9.00	9.3	1	Honancho
13	16905	17019	16982	16947	16027	16419	8.97	−2.9	1	(A) Nerima
14	13450	13408	13530	13266	12981	12504	9.00	−7.0	1	(A) Keisei-Tateishi
15	15315	15204	15384	15359	14579	14734	8.21	−3.8	1	(A) Kawaguchi
16	10085	10340	10291	10086	10091	10079	5.16	−0.1	1	(A) Kita-Urawa
17	7345	7432	7428	7294	7529	7639	7.73	4.0	1	Satte
18	13325	13277	13385	13092	13454	13890	9.37	4.2	1	Kanazawabunko
19	2085	2071	2101	2062	1978	1987	8.51	−4.7	1	
20	720	736	747	770	768	750	9.40	4.2	0	
21	325	330	326	335	336	328	8.19	0.9	0	
22	315	318	317	300	297	296	7.97	−6.0	0	
23	355	349	337	337	334	329	7.52	−7.3	0	
24	120	124	121	123	121	121	9.03	0.8	0	
25	355	341	344	341	340	338	6.58	−4.8	0	
Tokyo area	25176675	28303423	30124349	31829959	33456036	34286604	32.10	36.2		

Source: Reproduced by the author with data from the Population Census by the Statistics Bureau, Ministry of Internal Affairs and Communications.
Notes:
In the "75DID" column, 1 indicates inside DID and indicates outside DID.
In the "Station" column, (A) indicates the railway station in an adjacent grid square.

175

station, which indicates that they are conveniently located in terms of public transport service.

The 14 stable districts located in the Metropolis of Tokyo are shown in Figure 7.4. They are found in the ward area outside Loop Road No. 7 and inside Loop Road No. 8. No districts are located inside the JR Yamanote Line but the majority are concentrated to the west and southwest of the city centre (see Section 4 for a description of control districts A, B and C).

The following section examines the characteristics of those districts based on population census results.

3. Characteristics and typology of stable districts

3.1 Rate of change in age cohort

The main characteristic of stable districts is the rate of change in age cohort, as shown in Table 7.2, which is calculated as follows:

$$r(t, i) = p(t, i)/p(t - 5, i - 5),$$

where $i = p(t, i)$: population aged i to $i + 4$ in year t,

$i = 5, 10, 15, \ldots, 70, 75, 80$, $t = 1980, 1985, 1990, 1995$,

with $p(t, 80)$ indicating population aged 80 or over.

(Population data with quinquennial age brackets have been consistently gathered and available since 1980 only.)

Figure 7.5 shows average rate of change in age cohort for the whole Tokyo area and for each type of stable district, to be described later. Figure 7.6 shows population by age bracket in District Nos. 5, 8 and 23, which are most representative of individual types of stable district (Types I-1, I-2 and II (actually type II-2) respectively, according to the typology described below).

The rate of change in age cohort has been stable over the years in a majority of the districts. Typically, an upward trend in the population aged 15–24 is visible in many districts. They may be classified into the districts that experience a downward trend in population in the late 20s age bracket, and those that experience a decline a little later, in the early 30s or the late 30s bracket. For the purpose of this chapter, the former is called a Type I-1 district, the latter a Type I-2 district. Those districts

Table 7.2 Rate of change in age cohort in stable districts: r(t, i) 1980 to 1985

No.	Type	0–4 → 5–9	5–9 → 10–14	10–14 → 15–19	15–19 → 20–24	20–24 → 25–29	25–29 → 30–34	30–34 → 35–39	35–39 → 40–44	40–44 → 45–49	45–49 → 50–54	50–54 → 55–59	55–59 → 60–64	60–64 → 65–69	65–69 → 70–74	70–74 → 75–79	75–79 → 80–84	80– → 85–
1	I-1	0.85	0.92	1.18	1.56	0.83	0.81	0.84	0.92	0.98	0.94	0.94	0.93	0.92	0.85	0.80	0.70	0.47
2	I-1	0.85	0.96	1.28	1.81	0.81	0.87	0.89	0.97	1.01	0.98	0.96	0.97	0.93	0.91	0.81	0.80	0.55
3	I-1	0.84	0.91	1.28	1.78	0.77	0.87	0.84	0.88	0.95	0.91	0.94	0.90	0.89	0.89	0.79	0.74	0.55
4	I-1	0.93	1.02	1.30	1.94	0.74	0.79	0.85	0.99	1.00	0.99	0.93	0.90	0.93	0.86	0.85	0.71	0.50
5	I-1	0.89	1.02	1.23	1.72	0.66	0.82	0.87	0.91	0.89	0.91	0.88	0.92	0.90	0.86	0.80	0.73	0.46
6	I-1	0.91	0.93	1.05	1.31	0.82	0.92	0.90	0.95	1.00	0.96	0.94	0.93	0.95	0.97	0.87	0.70	0.51
7	I-1	0.93	0.91	1.19	1.87	0.78	0.82	0.90	0.88	0.96	0.98	0.96	0.96	0.89	0.90	0.85	0.71	0.54
9	I-1	0.88	0.97	1.19	1.25	0.83	0.85	0.85	0.94	0.98	0.98	0.93	0.94	0.90	0.86	0.76	0.66	0.57
10	I-1	0.86	0.93	1.22	1.47	0.80	0.86	0.90	0.93	0.99	0.91	0.92	0.97	0.95	0.90	0.85	0.66	0.46
11	I-1	0.84	0.97	1.27	1.80	0.67	0.72	0.85	0.91	1.00	1.00	0.95	0.89	0.88	0.88	0.84	0.75	0.44
12	I-1	0.86	0.93	1.17	1.58	0.86	0.89	0.90	0.95	0.98	0.99	0.94	0.92	0.90	0.87	0.75	0.75	0.51
13	I-1	0.81	0.91	1.12	1.58	0.84	0.88	0.83	0.91	0.96	0.95	0.89	0.94	0.92	0.93	0.82	0.69	0.56
19	I-1	0.90	0.92	1.25	1.78	0.60	0.74	0.96	0.93	0.91	0.88	0.93	0.95	0.99	0.88	0.88	0.89	0.42
8	I-2	0.72	0.86	0.99	1.17	1.04	1.03	0.82	0.89	0.89	0.90	0.96	0.90	0.81	0.89	0.79	0.68	0.46
15	I-2	0.91	0.96	1.00	1.07	0.97	0.89	0.94	0.97	0.95	0.96	0.93	0.91	0.88	0.85	0.70	0.63	0.51
16	I-2	0.83	0.90	1.08	1.22	0.92	0.82	0.84	0.90	0.99	0.93	0.91	0.92	0.91	0.87	0.85	0.72	0.48
18	I-2	0.81	0.76	0.94	1.17	1.20	1.05	0.81	0.80	0.88	0.96	0.97	0.93	0.89	0.77	0.89	0.67	0.52
14	II-1	0.94	0.97	0.95	0.94	0.93	0.88	0.97	0.98	0.95	0.93	0.94	0.95	0.90	0.85	0.75	0.72	0.54
17	II-1	1.06	1.02	1.05	0.87	0.70	0.93	1.01	1.01	0.97	0.94	0.94	0.90	0.90	0.87	0.80	0.66	0.45
20	II-2	1.04	0.98	1.00	0.94	1.05	1.02	0.88	1.26	0.96	1.12	0.97	0.98	0.95	0.85	0.88	0.60	0.71
21	II-2	1.14	1.06	1.00	0.81	1.07	1.00	0.97	1.00	1.06	1.00	0.97	0.96	1.00	0.94	0.70	0.42	0.33
22	II-2	0.94	0.93	0.86	0.74	0.77	1.17	0.89	0.95	1.00	1.04	0.90	0.84	0.95	0.78	0.79	0.40	0.40
23	II-2	1.00	1.04	0.93	0.87	0.94	0.93	0.96	0.96	1.04	0.93	1.00	0.94	1.00	0.82	1.00	0.90	0.20
24	II-2	0.83	1.00	0.60	0.78	1.00	1.21	1.14	1.00	1.20	1.00	1.00	1.00	0.83	1.00	0.50	1.00	0.20
25	II-2	1.05	1.10	0.84	0.62	0.76	1.13	1.32	0.88	1.00	1.00	1.00	0.96	1.00	1.09	0.50	0.70	0.43
Tokyo area		1.00	1.00	1.09	1.21	0.94	0.99	0.99	1.00	1.00	0.99	0.97	0.96	0.94	0.90	0.83	0.72	0.50

177

Table 7.2 (cont.)
1985 to 1990

No.	Type	0–4 / 5–9	5–9 / 10–14	10–14 / 15–19	15–19 / 20–24	20–24 / 25–29	25–29 / 30–34	30–34 / 35–39	35–39 / 40–44	40–44 / 45–49	45–49 / 50–54	50–54 / 55–59	55–59 / 60–64	60–64 / 65–69	65–69 / 70–74	70–74 / 75–79	75–79 / 80–84	80– / 85–
1	I-1	0.88	0.92	1.13	1.43	0.78	0.83	0.82	0.89	0.96	0.97	0.90	0.93	0.90	0.88	0.86	0.76	0.59
2	I-1	0.85	0.97	1.16	1.50	0.80	0.81	0.82	0.87	0.95	0.95	0.93	0.90	0.87	0.88	0.83	0.71	0.50
3	I-1	0.81	0.88	1.27	1.70	0.72	0.81	0.80	0.94	1.00	0.99	0.96	0.95	0.96	0.94	0.87	0.76	0.54
4	I-1	0.96	1.00	1.30	1.87	0.79	0.78	0.89	0.98	0.99	0.93	0.91	0.92	0.94	0.87	0.83	0.73	0.44
5	I-1	0.92	1.02	1.31	1.68	0.65	0.81	0.89	0.99	0.97	0.94	0.92	0.92	0.93	0.90	0.85	0.74	0.49
6	I-1	0.86	0.86	1.06	1.34	0.86	0.83	0.80	0.85	0.91	0.88	0.92	0.94	0.95	0.86	0.80	0.65	0.54
7	I-1	0.72	0.89	1.26	1.90	0.80	0.82	0.78	0.89	1.01	1.00	0.98	0.97	0.92	0.89	0.88	0.80	0.68
9	I-1	0.92	0.96	1.21	1.29	0.85	0.81	0.90	0.91	0.96	0.97	0.94	0.96	0.89	0.92	0.81	0.71	0.49
10	I-1	0.83	0.86	1.19	1.51	0.83	0.80	0.82	0.87	0.93	0.90	0.89	0.91	0.87	0.89	0.80	0.71	0.53
11	I-1	0.93	1.04	1.37	1.82	0.78	0.79	0.83	0.95	1.03	0.98	0.93	0.90	0.92	0.90	0.80	0.72	0.54
12	I-1	0.83	0.95	1.14	1.53	0.91	0.91	0.87	0.96	1.00	0.93	0.94	0.92	0.90	0.89	0.78	0.70	0.52
13	I-1	0.77	0.86	1.03	1.32	0.75	0.79	0.79	0.92	0.92	0.94	0.95	0.94	0.96	0.89	0.81	0.73	0.54
19	I-1	0.88	0.85	1.32	1.85	0.55	0.61	0.82	0.84	0.97	0.89	1.01	1.00	1.01	0.88	0.75	0.79	0.42
8	I-2	0.73	0.84	1.11	1.41	1.07	0.93	0.78	0.86	0.88	0.98	0.86	0.85	0.85	0.87	0.80	0.82	0.38
15	I-2	0.74	0.84	0.99	1.24	1.03	0.82	0.85	0.87	0.93	0.93	0.91	0.90	0.85	0.86	0.67	0.65	0.39
16	I-2	0.85	0.94	1.05	1.34	0.95	0.91	0.90	0.90	0.97	0.95	0.93	0.93	0.88	0.86	0.89	0.80	0.51
18	I-2	0.80	0.79	1.07	1.26	1.34	1.19	0.88	0.86	0.94	0.93	0.86	0.93	0.85	0.83	0.82	0.70	0.53
14	II-1	0.89	0.96	1.023	1.04	0.92	0.89	0.95	1.00	0.98	0.96	0.95	0.95	0.86	0.89	0.75	0.58	0.38
17	II-1	1.08	1.06	1.06	0.98	0.89	0.98	1.10	1.04	1.03	1.02	0.99	0.98	0.99	0.98	0.80	0.71	0.47
20	II-2	1.10	1.02	1.00	0.88	1.02	0.93	0.98	1.12	1.00	0.98	0.78	1.07	1.02	0.86	0.69	0.57	0.45
21	II-2	1.13	0.96	0.76	0.75	1.06	0.97	1.04	0.93	0.82	1.00	1.04	0.98	1.00	0.94	0.73	0.86	1.00
22	II-2	0.82	1.18	0.86	0.61	1.36	0.79	0.93	1.06	1.05	0.95	0.96	1.05	0.88	1.05	0.86	0.82	0.25
23	II-2	1.05	1.33	0.92	0.86	0.96	0.81	1.00	1.08	1.04	0.92	1.04	0.96	1.00	0.90	1.00	0.63	0.40
24	II-2	1.18	1.00	0.80	0.67	0.86	1.25	0.94	0.88	1.00	1.00	0.89	1.00	0.92	0.60	1.00	1.00	0.25
25	II-2	0.91	1.05	0.91	0.69	1.22	1.04	0.96	1.04	1.14	0.95	0.97	0.88	0.86	0.94	0.67	0.80	0.40
Tokyo area		1.00	1.01	1.08	1.18	0.97	1.00	1.00	1.00	1.00	0.99	0.98	0.96	0.94	0.91	0.85	0.74	0.52

178

1990 to 1995

No.	Type	0–4 → 5–9	5–9 → 10–14	10–14 → 15–19	15–19 → 20–24	20–24 → 25–29	25–29 → 30–34	30–34 → 35–39	35–39 → 40–44	40–44 → 45–49	45–49 → 50–54	50–54 → 55–59	55–59 → 60–64	60–64 → 65–69	65–69 → 70–74	70–74 → 75–79	75–79 → 80–84	80– → 85–
1	I-1	1.06	0.98	1.15	1.27	0.87	0.90	0.97	0.96	1.02	0.95	0.95	0.94	0.92	0.88	0.82	0.75	0.54
2	I-1	0.95	0.92	1.20	1.44	0.90	0.89	0.89	0.90	1.00	0.99	0.96	0.94	0.97	0.89	0.82	0.72	0.51
3	I-1	0.88	0.96	1.13	1.48	0.79	0.88	0.85	0.91	0.92	0.94	0.91	0.91	0.92	0.93	0.83	0.77	0.53
4	I-1	0.97	1.09	1.27	1.62	0.88	0.83	0.87	0.93	0.94	0.94	0.93	0.94	0.94	0.90	0.87	0.68	0.50
5	I-1	0.93	1.01	1.39	1.85	0.81	0.84	0.90	0.94	0.98	0.93	1.00	0.96	0.91	0.86	0.86	0.77	0.55
6	I-1	0.86	0.95	1.60	1.27	0.96	0.96	0.91	0.94	0.97	0.95	0.99	0.98	0.97	0.92	0.87	0.81	0.59
7	I-1	0.86	0.95	1.33	1.52	0.84	0.85	0.81	0.94	1.01	1.01	0.99	0.97	0.92	0.88	0.87	0.77	0.53
9	I-1	0.91	1.04	1.34	1.28	0.81	0.86	0.91	0.97	1.09	1.01	0.96	0.94	0.95	0.96	0.90	0.77	0.53
10	I-1	0.97	1.00	1.28	1.47	0.81	0.80	0.86	0.93	0.98	0.99	0.96	0.97	0.94	0.97	0.88	0.78	0.58
11	I-1	0.97	1.06	1.34	1.54	0.77	0.73	0.89	0.92	1.02	1.00	1.01	0.93	0.92	0.90	0.85	0.74	0.58
12	I-1	0.87	0.99	1.16	1.40	0.95	0.88	0.83	0.94	0.97	0.97	1.00	0.92	0.93	0.92	0.92	0.81	0.63
13	I-1	0.95	0.96	1.12	1.40	0.97	0.92	0.96	1.01	1.01	0.98	0.98	0.97	0.92	0.93	0.80	0.76	0.45
19	I-1	1.13	1.01	1.08	1.57	0.62	0.76	0.95	1.06	1.05	1.04	1.07	0.92	0.98	0.91	0.92	0.82	0.64
8	I-2	0.63	0.74	1.25	1.41	1.11	0.97	0.71	0.77	0.88	0.99	0.98	0.94	0.91	0.89	0.85	0.76	0.53
15	I-2	0.80	0.92	1.03	1.18	1.10	0.87	0.92	0.97	1.03	1.01	1.00	0.96	0.94	0.93	0.80	0.73	0.52
16	I-2	0.91	0.96	0.93	1.16	1.09	0.98	0.96	0.91	0.91	0.96	0.95	0.94	0.92	0.94	0.86	0.72	0.54
18	I-2	0.82	0.91	1.02	1.27	1.07	1.08	0.88	0.89	1.01	1.01	0.99	0.98	0.97	0.95	0.91	0.76	0.57
14	II-1	1.02	0.89	0.98	0.93	0.89	1.00	1.04	0.98	0.93	0.95	0.93	0.88	0.87	0.86	0.75	0.71	0.45
17	II-1	1.12	1.07	1.04	0.97	0.91	0.96	1.10	1.05	1.03	0.99	0.98	0.96	0.90	0.86	0.85	0.77	0.62
20	II-2	1.07	1.17	1.06	0.63	0.80	0.90	1.05	1.11	1.05	1.05	1.09	1.09	0.95	0.80	0.93	0.70	0.40
21	II-2	1.14	1.00	0.88	0.85	1.67	1.06	1.00	0.96	1.04	1.00	0.94	1.00	0.97	0.74	0.93	0.82	0.46
22	II-2	1.13	1.07	0.90	0.92	1.45	0.79	0.84	1.08	1.39	1.05	0.95	0.96	0.95	0.93	0.75	0.92	0.60
23	II-2	1.00	1.05	0.94	0.83	1.04	1.00	1.00	1.07	1.00	1.00	1.00	0.96	0.94	1.00	1.00	0.71	0.44
24	II-2	1.00	0.85	1.00	0.88	2.50	0.67	0.80	1.00	1.14	1.00	1.00	0.75	1.00	0.67	0.67	1.00	0.60
25	II-2	1.17	1.00	0.95	0.90	0.78	0.82	0.93	0.88	0.96	1.00	1.17	0.94	0.95	0.94	0.76	0.75	0.25
Tokyo area		0.99	0.99	1.07	1.12	0.96	0.98	0.98	0.99	1.00	0.99	0.98	0.96	0.94	0.91	0.86	0.76	0.54

Source: Reproduced by the author with data from the Population Census by the Statistics Bureau, Ministry of Internal Affairs and Communications.

179

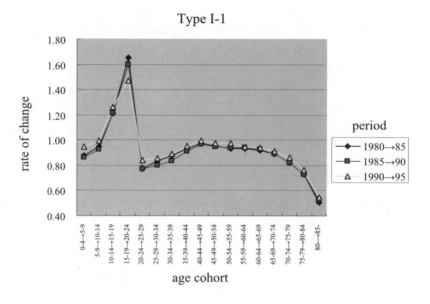

Type I-1

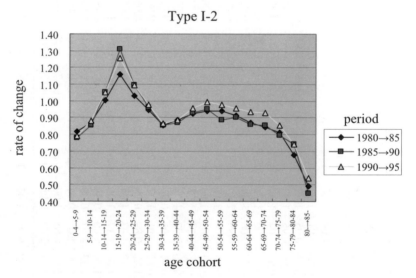

Type I-2

Figure 7.5 Average rate of change in age cohort by type of district (*Source*: Produced by the author using data from Population Census by the Statistics Bureau, Ministry of Internal Affairs and Communications)

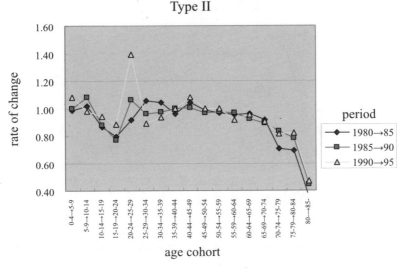

Figure 7.5 (cont.)

are located in grid squares with a population density of 10,000/km^2 or greater.

In those stable districts that do not show an upward trend in population in the 15–19 or 20–24 age bracket, the rate of change in age cohort remains at around 1.00 for all age brackets. Although population density in most of such districts is below 1,000/km^2, some DIDs are also included in this category. For the purpose of this chapter, the DIDs in this category are classified as a Type II-1 district, while the others are defined as Type II-2. The typology is mentioned in Table 7.2.

By age bracket, the share of young people has always been largest in Type I districts (No. 5), whereas in Type II districts (No. 23), the demographic structure moves almost in line with the ageing of the population (Figure 7.6).

Looking at geographical distribution, Type I-1 districts are located in the ward area outside Loop Road No. 7 and along the JR Chuo Line. Type II-2 districts are found as far as 50 km from the city centre. Type I-2 and Type II-1 districts are generally found between those two types of district, mainly outside the Metropolis of Tokyo.

3.2 Education/occupation status

Table 7.3 shows the composition of population by education/occupation status. In many Type I districts, junior college, technical college, college

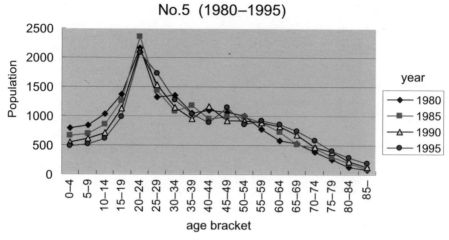

Population by age bracket in District No. 5

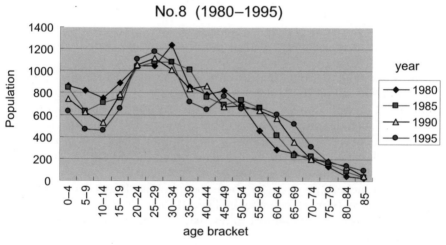

Population by age bracket in District No. 8

Figure 7.6 Population by age bracket in districts Nos. 5, 8 and 23 (*Source*: Produced by author using data from Population Census by the Statistics Bureau, Ministry of Internal Affairs and Communications)

and graduate students make up a substantial share of the population, indicating that students largely account for the inflow of young people over the years. In stable districts within the DID area, the share of tertiary

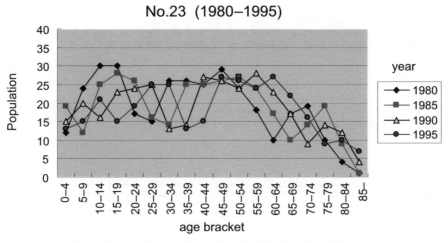

Population by age bracket in District No. 23

Figure 7.6 (cont.)

sector employees was already high in the 1970s. This implies that those districts were among the first to experience urbanization in the metropolitan area. In a few stable districts, secondary sector employees have a relatively large share of the population (Nos. 14 and 15).

3.3 Duration of residency and previous residence

Table 7.4 shows the duration of residency in the stable districts and previous residence. Immigration rate in the recent five years is relatively high in many Type I districts, pointing to substantial movements in population. The majority of the population in Type II-2 districts has lived there since birth, indicating low population mobility. In Type II-1 districts, population mobility is relatively low but shows some activity. The 1980 Census data indicates that a relatively high percentage of people have lived in many of the stable districts since before 1965. It can thus be understood that transient residents mingle with long-term residents in stable districts within the DID boundary.

According to the 1990 Census, Tokyo Metropolis districts that are relatively remote from the city centre experienced substantial migration from other cities in the Metropolis, whereas migrants from other prefectures predominate in districts outside the Metropolis of Tokyo. Migrants from inside the city have a considerable share in District No. 14.

Table 7.3 Population by education/occupation status

No.	1970 College students*	1995 College students*	1970 Primary sector	1970 Secondary sector	1970 Tertiary sector	1995 Primary sector	1995 Secondary sector	1995 Tertiary sector
1	7.0	6.2	0.8	33.5	65.7	0.6	17.6	79.5
2	6.5	6.3	0.3	30.9	68.8	0.2	18.6	79.5
3	10.3	8.4	0.6	25.3	73.8	0.2	16.6	81.1
4	9.9	8.1	0.8	22.7	76.1	0.2	16.3	81.8
5	8.1	9.2	0.5	24.5	74.3	0.3	17.7	79.7
6	4.1	4.9	0.1	33.5	66.5	0.1	22.0	76.8
7	8.7	8.3	0.5	31.4	68.0	0.5	19.0	79.3
8	3.2	5.7	1.4	28.7	69.8	0.3	21.4	76.6
9	4.8	5.8	1.1	37.3	61.5	0.6	24.7	73.0
10	8.0	6.2	0.3	28.4	71.0	0.5	19.1	77.4
11	8.3	8.6	0.9	33.6	65.2	0.1	19.0	78.3
12	5.9	5.3	0.3	32.5	66.9	0.2	19.6	77.1
13	5.6	5.0	0.3	28.5	71.0	0.3	20.6	76.9
14	1.7	3.0	0.0	59.5	40.1	0.0	40.6	57.6
15	2.3	3.3	0.0	56.3	43.7	0.1	34.1	63.9
16	3.7	5.4	1.1	32.6	66.3	0.4	22.8	76.3
17	1.6	3.3	3.8	39.3	57.0	0.6	35.8	63.3
18	4.1	4.4	0.3	33.9	65.8	0.2	23.4	75.7
19	0.2	0.1	8.1	16.2	75.6	2.6	13.9	83.5
20	0.0	1.2	23.8	36.9	39.3	8.4	33.4	58.2
21	0.0	0.3	63.4	9.8	26.8	22.4	30.5	47.1
22	0.0	0.7	80.6	8.3	11.1	16.9	34.4	48.8
23	0.0	0.6	58.1	11.6	30.2	9.6	32.5	57.2
24	0.0	0.0	76.5	23.5	0.0	24.2	37.9	37.9
25	0.0	0.3	51.4	21.6	27.0	10.7	35.3	54.0
Tokyo area	3.4	4.5	8.8	38.6	52.5	2.3	29.4	67.1

Source: Reproduced by the author with data from the Population Census by the Statistics Bureau, Ministry of Internal Affairs and Communications.

Notes:

Enrolled students ÷ population × 100, workers in each sector ÷ total workers × 100.

Junior college, technical college, college and graduate students.

3.4 Commuting

Table 7.5 shows the means of transportation to workplace or school. In stable districts, the general trend is that railway is the principal means of transportation in the DID area while motorists predominate outside the DID boundary.

Short-distance commuting of 0–29 minutes is rare in many Type I-1 districts. The substantial share of middle-to-long-distance commuting in 30–90 minutes implies that those districts are residential areas mostly inhabited by company employees working in the city centre. In contrast, most people seem to be employed locally in Type II districts, as suggested by the large share of short-distance commuting. Commuting distance is evenly distributed in Type I-2 districts, pointing to a mixture of those working in the city centre and local employees.

As suggested by the predominance of the secondary sector, District Nos. 14 and 15 are mainly home to blue-collar workers employed locally.

3.5 Housing

Tables 7.6 and 7.7 show the composition of dwellings by ownership and by type of building, respectively. Type I-1 districts have a large proportion of apartment houses for rent, indicating the early development of a housing stock to accommodate young transient dwellers. The low population mobility in Type II districts may be explained by the substantial share of owner-occupied houses. The considerable share of company houses in many Type I-2 districts seems to be explained by the constant inflow of workers in their late twenties. From the composition by floor area, shown in Table 7.8, a certain amount of diversity in the scale of dwellings is also observed within the DID area.

3.6 Summary and physical pattern

Thus, many of the stable districts have a population density of over 10,000/km^2, having experienced a substantial inflow of young people over the years. This inflow may be classified into two types, students (within the Metropolis of Tokyo) and young workers living in company houses (mainly outside the Metropolis). Stable home-ownership rates are observed in those districts, as well as the diversity of dwellings in terms of scale. The large share of tertiary sector workers, observed as early as 1970, and the predominance of medium–long-distance commuting implies that the districts developed as residential areas for company employees working in the city centre. That is why various types of households coexist in the districts. Short-term dwellers mingle with a consider-

Table 7.4 Duration of residency in stable districts and previous residence

No.	1970 Census		1980 Census					Since before 1985	Since 1985–90	1990 Census		
	Since birth	Since 1965–70	Since birth	Since before 1964	Since 1965–69	Since 1970–75	Since 1975–80			Inside the city (see Notes)	Another city in the prefecture (see Notes)	Outside the prefecture (see Notes)
1	16.1	59.5	14.8	21.0	8.0	17.0	39.2	65.4	30.8	32.15	25.18	42.67
2	16.1	63.8	13.7	21.1	7.1	15.4	42.4	60.7	34.7	23.24	32.95	43.82
3	11.6	72.4	11.5	19.0	7.3	15.9	46.1	58.6	36.8	28.71	24.92	46.37
4	14.8	67.1	12.5	22.0	7.1	14.8	43.4	59.5	36.8	29.06	28.23	42.71
5	13.9	66.8	11.9	19.7	11.3	15.9	41.0	59.7	36.3	19.61	31.33	49.06
6	17.3	49.2	14.4	23.0	6.7	19.2	36.4	68.4	25.3	23.91	40.90	35.19
7	17.4	65.1	14.3	17.7	8.2	15.0	44.4	56.2	38.7	19.71	38.16	42.13
8	14.7	63.6	13.7	17.8	11.3	16.7	40.3	57.3	34.8	14.95	37.84	47.21
9	16.1	54.8	14.0	23.6	9.1	15.2	38.0	66.3	28.5	21.41	41.66	36.93
10	11.9	71.6	11.1	17.9	8.0	16.5	46.2	59.5	35.1	27.78	31.61	40.61
11	14.1	63.7	13.2	23.8	8.7	14.6	39.5	60.8	35.3	21.09	34.66	44.26
12	13.7	71.5	12.0	16.8	9.4	17.0	44.4	60.5	34.2	27.38	32.76	39.86
13	14.1	66.5	13.4	19.5	6.1	16.1	44.8	63.3	30.9	31.98	26.12	41.90
14	19.1	57.8	18.4	21.6	11.8	19.3	28.8	74.9	20.8	51.63	24.24	24.13
15	22.7	53.1	20.0	21.3	9.2	15.4	33.8	68.1	26.7	34.80	16.29	48.91
16	18.9	57.1	17.1	21.6	7.7	17.6	35.9	64.0	30.3	19.61	20.46	59.92
17	33.0	44.7	28.0	22.7	11.1	14.3	23.8	75.9	20.1	32.89	30.12	36.99
18	20.8	69.4	17.1	18.0	6.6	19.2	39.0	60.7	32.1	27.34	26.20	46.46
19	25.7	70.3	20.8	17.1	8.7	14.2	39.2	63.2	33.9	20.42	14.61	64.98
20	50.7	14.6	44.2	25.6	7.2	10.7	12.3	81.9	12.8	20.41	27.55	52.04
21	61.5	3.1	57.7	28.5	2.1	4.9	6.7	91.7	4.2	7.14	92.86	0.00
22	65.1	3.2	63.7	25.6	1.9	2.2	6.6	89.9	5.1	33.33	53.33	13.33
23	57.7	9.9	53.4	26.4	8.9	6.2	5.0	90.7	4.8	0.00	62.50	37.50

24	62.5	8.3	58.7	29.8	1.7	2.5	7.4	90.1	3.3	0.00	75.00	25.00
25	63.4	7.0	49.4	21.2	6.7	9.0	13.7	83.5	9.7	51.52	30.30	18.18
Tokyo area	23.7	59.0	18.7	15.7	9.2	19.4	36.8	66.9	27.7	28.98	28.37	42.65

Source: Reproduced by the author with the data of Population Census by the Statistics Bureau, Ministry of Internal Affairs and Communications.
Notes:
Migrants in each period ÷ population × 100.
Migrants in each category ÷ migrants in 1985–90 × 100.

Table 7.5 Transportation to workplace or school

No.	1970 Census					1980 Census					1990 Census			
	Walk only	Railway	Bus	Car	Motor	Walk only	Railway	Bus	Car	Motor	0–29 min.	30–59 min.	60–89 min.	90 min.–
1	11.5	66.6	7.3	11.9	1.6	10.6	88.6	7.9	11.5	8.1	26.9	48.4	20.0	4.6
2	12.6	66.2	7.8	10.6	0.7	10.6	88.2	8.1	9.5	7.8	25.1	56.1	14.9	3.9
3	10.6	73.2	3.9	9.9	2.0	6.8	93.3	5.4	9.5	13.5	22.5	49.8	23.3	4.3
4	9.5	75.8	3.1	9.5	1.2	8.6	95.6	6.9	8.4	8.4	24.5	54.9	16.7	3.9
5	9.3	78.8	1.6	7.9	1.9	8.6	95.6	4.8	8.3	10.9	23.8	52.4	19.8	4.1
6	18.9	52.8	9.7	10.0	7.9	9.3	60.4	16.0	15.8	36.5	38.2	23.5	26.9	11.3
7	13.9	68.6	5.4	6.2	5.6	7.9	86.0	11.4	7.3	23.7	28.7	34.6	31.7	5.0
8	10.6	67.8	5.3	12.6	3.0	7.3	89.4	12.9	11.3	17.3	24.1	27.3	40.2	8.4
9	14.1	63.3	9.1	8.8	4.5	8.8	81.4	13.5	11.0	21.3	29.2	30.2	33.9	6.6
10	7.6	73.3	5.5	9.7	3.0	7.9	92.8	15.7	9.6	15.0	23.9	40.3	30.0	5.7
11	10.7	74.9	3.0	8.8	2.0	7.8	94.7	7.0	9.2	10.5	22.5	52.9	20.1	4.5
12	12.7	64.1	6.9	12.4	2.7	9.8	83.1	10.3	8.9	15.7	27.2	49.9	18.7	4.1
13	12.5	63.1	9.0	9.8	4.2	10.9	78.7	18.5	10.2	21.4	30.1	32.6	31.3	6.0
14	31.9	43.8	4.1	11.6	8.4	11.7	58.4	8.0	11.0	34.4	40.4	35.1	19.7	4.7
15	21.4	44.7	7.0	10.5	15.6	10.1	60.3	11.2	14.7	31.0	37.7	27.1	29.2	5.9
16	14.0	62.7	8.5	9.9	4.7	8.2	71.5	9.8	14.5	22.4	32.9	19.0	38.7	9.4
17	17.1	56.4	1.7	10.2	13.9	9.3	49.6	2.5	27.8	22.3	45.7	16.6	17.7	19.9
18	15.4	66.4	6.9	7.4	3.5	9.5	82.7	12.0	12.3	13.8	29.7	26.6	26.7	17.0
19	56.3	11.1	15.9	6.3	10.1	43.5	5.8	4.7	30.3	19.6	87.9	5.6	1.5	4.9
20	33.3	11.1	11.1	20.4	24.1	4.3	16.7	13.0	58.7	19.7	48.7	25.7	13.7	12.0
21	5.9	11.8	11.8	17.6	52.9	7.4	7.4	0.0	71.3	10.2	54.6	28.7	7.4	9.3
22	12.5	12.5	12.5	25.0	37.5	0.0	4.9	8.2	82.0	7.4	47.5	33.6	13.1	5.7
23	13.3	0.0	40.0	26.7	20.0	2.9	2.9	11.1	68.4	17.0	63.7	19.9	11.7	4.7
24	0.0	0.0	16.7	50.0	33.3	3.9	0.0	3.9	70.6	13.7	72.5	23.5	2.0	2.0
25	15.8	10.5	5.3	21.1	47.4	3.1	9.2	0.0	74.8	22.1	74.0	12.2	9.9	3.8

| Tokyo area | 18.9 | 51.2 | 9.8 | 12.4 | 7.0 | 9.5 | 61.9 | 15.3 | 23.8 | 21.8 | 37.8 | 28.0 | 23.0 | 11.3 |

Source: Reproduced by the author with data from the Population Census by the Statistics Bureau, Ministry of Internal Affairs and Communications.

Notes:

*People using each means of transportation ÷ commuters × 100.

*People in each category ÷ commuters × 100.

"Motor" indicates those using a car or motorcycle.

Table 7.6 Composition of private households in the stable districts by ownership

No.	1970 Owner-occupied	1970 Public rental	1970 Private rental	1970 Company houses	1970 Lodging	1995 Owner-occupied	1995 Public rental	1995 Private rental	1995 Company houses	1995 Lodging
1	40.5	0.6	46.3	9.3	3.3	42.2	0.0	51.2	5.1	1.5
2	42.0	1.0	46.0	8.0	3.1	40.5	0.3	51.2	6.7	1.4
3	34.4	4.2	45.0	12.0	4.4	37.2	2.7	51.9	6.6	1.7
4	36.9	2.3	50.8	6.8	3.1	34.5	1.5	57.7	4.7	1.5
5	37.3	11.4	39.5	9.6	2.2	35.8	7.2	48.4	7.1	1.6
6	28.4	28.8	36.0	5.1	1.7	34.3	24.9	34.2	5.8	0.7
7	35.6	2.7	52.5	8.1	1.2	37.9	1.8	52.3	6.7	1.2
8	35.2	3.0	39.6	21.1	1.1	37.0	1.0	47.2	13.5	1.3
9	41.3	14.2	38.4	5.0	1.1	41.5	13.5	39.1	4.4	1.5
10	33.3	5.9	51.2	7.7	1.9	34.5	3.9	53.7	6.3	1.5
11	45.3	1.5	43.9	6.2	3.1	41.3	0.0	52.3	3.9	2.5
12	37.2	4.0	48.1	8.7	2.0	40.8	3.2	48.6	6.1	1.3
13	31.3	7.3	48.6	9.2	3.6	38.7	4.3	48.4	7.0	1.6
14	48.8	6.3	36.3	7.7	1.0	61.9	4.1	29.9	2.8	1.4
15	41.2	4.4	47.0	5.8	1.6	44.0	4.1	46.0	4.5	1.5
16	47.6	0.9	33.9	16.7	0.9	46.7	1.0	37.5	14.1	0.7
17	63.2	0.6	32.9	3.0	0.3	67.4	0.0	29.3	2.5	0.8
18	33.3	8.9	39.0	16.5	2.3	41.9	6.4	26.2	14.1	1.4
19	68.5	2.2	23.6	4.5	1.1	64.2	0.0	27.8	7.1	0.9
20	90.6	0.0	6.3	3.1	0.0	91.0	2.7	1.3	4.0	0.9
21	100.0	0.0	0.0	0.0	0.0	100.0	0.0	0.0	0.0	0.0
22	100.0	0.0	0.0	0.0	0.0	97.1	0.0	1.4	0.0	1.4
23	100.0	0.0	0.0	0.0	0.0	100.0	0.0	0.0	0.0	0.0
24	100.0	0.0	0.0	0.0	0.0	100.0	0.0	0.0	0.0	0.0
25	85.7	0.0	7.1	7.1	0.0	95.6	0.0	1.1	2.2	1.1
Tokyo area	47.5	7.4	36.2	7.2	1.6	52.7	7.1	33.5	5.7	1.1

Source: Reproduced by the author with data from the Population Census by the Statistics Bureau, Ministry of Internal Affairs and Communications.

Note:

*Households in each category of ownership ÷ private households × 100.

Table 7.7 Composition of private households in stable districts by type of building

No.	1980			1995					
	Single-dwelling	Row house	Apartment	Single-dwelling	Row house	Apartment	Apt. (1–2 stories)	Apt. (3–5 stories)	Apt. (6+ stories)
1	41.0	3.9	54.8	35.7	3.5	60.4	29.7	24.1	6.7
2	34.4	2.3	62.9	27.2	2.4	70.1	25.6	26.3	18.2
3	35.8	2.0	61.7	32.0	1.9	65.9	33.2	28.4	4.4
4	37.8	1.8	59.2	28.8	2.4	68.6	33.0	31.4	4.1
5	30.9	2.1	66.8	26.3	1.9	71.7	31.1	35.7	4.9
6	37.4	8.1	54.1	30.2	1.6	68.1	21.4	36.0	10.7
7	32.2	4.7	62.4	27.3	2.0	70.8	33.1	21.2	16.4
8	44.9	5.0	49.5	35.3	1.8	62.9	25.5	28.9	8.6
9	48.0	4.7	47.0	41.8	1.8	56.3	29.5	21.3	5.6
10	34.3	4.2	61.1	32.5	2.7	64.6	30.6	30.9	3.1
11	47.1	2.3	50.2	40.2	3.7	55.8	38.9	15.0	1.9
12	30.0	2.1	67.4	24.9	1.6	73.3	22.4	31.9	19.0
13	30.8	5.4	63.5	26.8	3.4	69.7	23.0	33.4	13.2
14	59.2	5.5	35.0	50.6	1.2	48.0	13.6	25.5	8.9
15	45.1	6.6	47.7	37.4	2.5	59.8	23.4	26.9	9.5
16	53.5	4.3	41.8	48.1	2.4	49.2	25.9	21.7	1.6
17	82.3	11.8	5.8	76.3	4.7	19.0	11.6	4.8	2.6
18	42.0	5.1	52.5	34.5	2.9	62.4	24.9	22.3	15.2
19	92.5	5.9	1.4	76.3	3.7	19.8	16.6	3.3	0.0
20	95.2	2.1	0.5	91.5	3.1	5.4	2.2	0.0	3.1
21	100.0	0.0	0.0	100.0	0.0	0.0	0.0	0.0	0.0
22	100.0	0.0	0.0	98.6	0.0	0.0	0.0	0.0	0.0
23	100.0	0.0	0.0	100.0	0.0	0.0	0.0	0.0	0.0
24	100.0	0.0	0.0	100.0	0.0	0.0	0.0	0.0	0.0
25	97.4	2.6	0.0	98.9	1.1	0.0	0.0	0.0	0.0
Tokyo area	54.2	4.6	40.7	47.1	2.5	50.2	18.0	21.2	11.0

Source: Reproduced by the author with data from the Population Census by the Statistics Bureau, Ministry of Internal Affairs and Communications.

Note:

*Households in each type of building ÷ private households × 100.

191

Table 7.8 Composition of private households in stable districts by floor area 1995

No.	0–29 m²	30–49 m²	50–69 m²	70–99 m²	100–149 m²	150 m²–
1	35.2	16.9	13.8	13.1	13.4	7.6
2	33.2	20.9	14.5	13.4	11.1	6.9
3	35.8	18.5	13.4	12.4	13.5	6.4
4	43.7	17.2	10.3	10.8	11.5	6.4
5	38.0	17.8	18.1	12.3	9.2	4.5
6	28.4	28.2	19.8	12.2	8.6	2.7
7	40.0	18.7	15.8	13.0	9.0	3.6
8	31.8	23.3	17.9	13.1	10.5	3.4
9	30.0	18.9	17.6	15.6	13.6	4.3
10	37.1	21.1	13.7	12.5	10.6	5.0
11	39.3	13.3	10.9	13.7	14.8	8.1
12	33.9	24.4	20.6	11.3	6.8	2.9
13	32.4	24.1	17.1	13.3	8.7	4.3
14	21.5	25.2	28.5	14.6	7.4	2.9
15	32.5	24.5	18.3	13.1	8.1	3.5
16	19.9	19.7	16.5	18.3	17.9	7.8
17	10.2	19.9	12.0	26.8	22.3	8.8
18	23.3	31.8	13.7	13.5	13.7	4.1
19	10.5	24.7	9.8	21.2	27.0	6.8
20	2.7	7.2	11.7	19.3	34.1	25.1
21	0.0	2.8	2.8	16.7	47.2	30.6
22	1.4	2.9	1.4	11.4	28.6	54.3
23	0.0	1.4	5.4	8.1	31.1	54.1
24	0.0	0.0	4.0	12.0	16.0	68.0
25	0.0	5.5	6.6	25.3	26.4	36.3
Tokyo area	20.2	20.6	18.1	19.4	15.6	6.1

Source: Reproduced by the author with data from the Population Census by the Statistics Bureau, Ministry of Internal Affairs and Communications.
Note:
*Households in each category of floor area ÷ private households × 100.

able number of long-term residents, who have been living there since before 1965.

An observation of land use in some Type I districts further clarifies the situation. Figures 7.7, 7.8 and 7.9 show land use maps based on City Planning Geographical Information System data obtained from the Tokyo Metropolis City Planning Bureau (as of 1996 for the 23 wards and 1997 for the Tama region). The maps are overlaid with approximate building polygons. Single- and multiple-dwelling houses have an overwhelming share in all three districts, thus ensuring residential function, but they

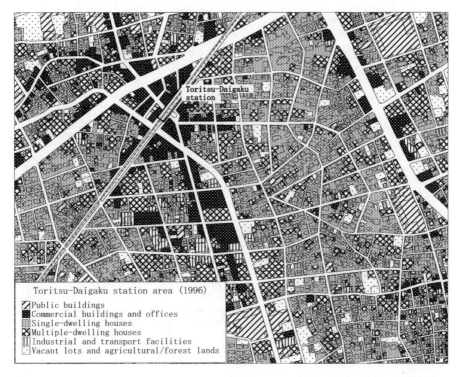

Figure 7.7 Land use map of District No. 2 (*Source*: Produced by the author using data from Geographical Information System by the Urban Planning Bureau, Tokyo Metropolitan Government)

are mixed with a considerable number of commercial sites. An analysis of chronological data (related maps are omitted due to space limitations) points to the stability, with a slight upward trend, of the space occupied by single-dwelling houses in terms of absolute floor area, although the districts share the general trend found in the ward area of a decrease in land for single-dwelling houses and an increase in land for multiple-dwelling houses.

In contrast, the stable districts outside the DID boundary (Type II-2), with population densities of less than 1,000/km^2, may be considered as rural districts with few urban characteristics, being located almost 50 km from the city centre. With little movement in population, those districts have experienced constant shifts in their demographic structures. It is not certain whether they remain stable when observed over a longer time span.

Significant inflow of young people is not observed in two of the stable

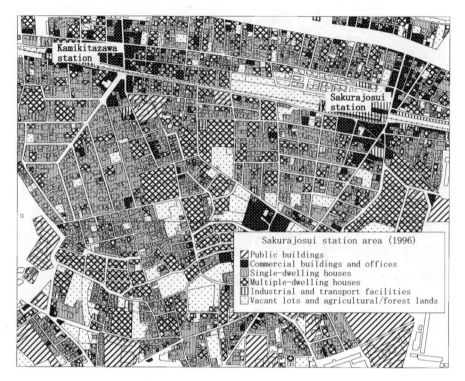

Figure 7.8 Land use map of District No. 5 (*Source*: Produced by the author using data from Geographical Information System by the Urban Planning Bureau, Tokyo Metropolitan Government)

districts located within the DID boundary (Type II-1): District Nos. 14 and 17, although the diversity of dwellings in scale and a certain extent of population movement may be confirmed. In both districts, a relatively small share of single population and a relatively large share of secondary sector employees with shorter commuting distances point to the existence of some industrial concentrations in their vicinity.

A constant and gradual inflow of population in the 30–39 and 0–9 age brackets is observed in District No. 17, indicating an inflow of families. Apparently, the district has experienced gradual development without large-scale projects, since its original development as a single-dwelling residential area at an early stage. The high proportion of intra-city migration in District No. 14 implies that residents in the district have moved into nearby houses of a different scale due to changing family membership or economic status.

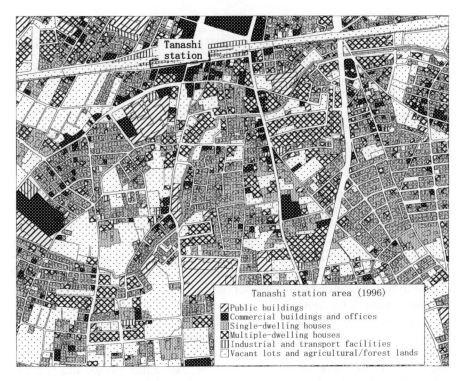

Figure 7.9 Land use map of District No. 9 (*Source*: Produced by the author using data from Geographical Information System by the Urban Planning Bureau, Tokyo Metropolitan Government)

4. Comparison with other districts

This section presents a comparison of the stable districts with other typical districts in the Metropolis of Tokyo. Districts A, B and C, as shown in Figure 7.4, have been chosen as control districts. Table 7.9 shows the population trends of the three districts. Table 7.10 shows the rate of change in age cohort over the years. Figures 7.10, 7.11 and 7.12 are graphs showing population by age bracket. The composition of private households is shown in Table 7.11 by ownership, and in Table 7.12 by floor area of dwellings.

District A has the largest absDPsum (= 90.4) throughout the 1970–1995 period for those grid squares with a population of 20,000 or over. Thus, it may be defined as "the district that experienced the most rapid change in population among the intensive urban districts". The district is located to the northwest of the JR Ikebukuro station and forms part of

Table 7.9 Population trends in control districts

No.	1970	1975	1980	1985	1990	1995	absDPsum	DP
A	31685	27566	34045	33234	21967	24349	90.04	−23.15
B	75	14396	15647	15163	15174	13542	19117.28	17956.00
C	1430	4126	6059	6850	8518	8553	273.20	498.11

Source: Reproduced by the author with the data of Population Census by the Statistics Bureau, Ministry of Internal Affairs and Communications.

the area formerly known as the "cheap lodgings belt". Although its demographic structure, shown in Figure 7.10, is similar to that of a stable district, a downward trend in the total population has been observed since 1980 (with a slight recovery in 1995). The increase in young population has not been as consistent as in the stable districts. As a matter of fact, Table 7.10 reveals that the late-teen population flowed out of the district during the bubble economy period (from 1985 to 1990). The increase in home ownership rate observed in Table 7.11 implies an outflow of house-renters, but household membership is generally smaller than in the stable districts, as can be seen in Table 7.12. District A also differs from the stable districts in that the floor area of single-dwelling houses has been declining in absolute terms, as well as in share (relevant data has been omitted due to space limitations). In other words, the rise in home ownership may be primarily attributed to an increase in condominiums. The land use map in Figure 7.13 shows an image of an intensive urban area, with a complicated mixture of various small-scale housing lots.

District B largely corresponds to the Suwa/Nagayama area in Tama New Town. Included in the first phase of the Tama New Town development project, the district saw a rapid increase in population from 1971, when people began to move in. Subsequently, however, the population decreased as early as 1980. Its demographic structure has two peaks, which correspond to baby-boomers (initial inhabitants) and their children. In contrast to the stable districts, however, an outflow of the younger population has become clear in recent years. Typically, households renting medium-scale (30–69 m^2) dwellings have come to account for a majority of the population. The land use map in Figure 7.14 shows that the district is largely made up of apartment houses.

District C, which lies adjacent to District B, covers part of the housing complex developed in the 1960s prior to the Tama New Town project and a housing complex developed after the project in the Suwa/Nagayama district, as well as the urban area, which has developed progressively along the New Town Avenue through land readjustments. In contrast to District B, the population of the districts has been increasing

Table 7.10 Rate of change in age cohort in control districts
1980 to 1985

No.	5–9	10–14	15–19	20–24	25–29	30–34	35–39	40–44
A	−17.00	−7.09	12.42	50.32	−26.75	−20.80	−14.66	−7.82
B	−20.44	−30.54	9.48	−0.97	136.58	41.28	−18.30	−28.87
C	−6.23	−16.20	34.77	70.57	48.88	36.62	−3.03	−11.51

No.	45–49	50–54	55–59	60–64	65–69	70–74	75–79	80–84
A	−7.72	−7.48	−9.03	−10.63	−12.29	−19.43	−24.71	−30.34
B	−19.96	−8.72	1.73	8.56	6.90	−15.08	−11.24	−32.43
C	−7.57	−0.28	−3.69	12.00	8.04	9.33	−16.67	8.89

1985 to 1990

No.	5–9	10–14	15–19	20–24	25–29	30–34	35–39	40–44
A	−44.20	−36.6	−21.13	4.69	−46.45	−45.38	−45.42	−40.58
B	−17.35	−18.05	−10.86	−14.24	40.33	14.16	−10.96	−14.26
C	3.14	1.01	14.35	71.53	34.20	30.96	12.20	2.23

No.	45–49	50–54	55–59	60–64	65–69	70–74	75–79	80–84
A	−32.72	−34.18	−36.63	−37.80	−42.02	−42.13	−43.92	−53.47
B	−8.60	−4.43	−7.72	−0.24	3.32	1.29	−10.28	−29.11
C	0.31	3.62	5.29	4.78	−2.14	−6.61	0.00	−5.00

1990 to 1995

No.	5–9	10–14	15–19	20–24	25–29	30–34	35–39	40–44
A	−3.39	3.78	29.96	66.80	0.34	−4.53	−1.22	9.36
B	−29.20	−26.93	−15.72	−20.55	−4.62	12.51	−26.30	−25.06
C	−25.95	−19.43	16.33	34.61	−3.32	−8.19	−17.77	−11.70

No.	45–49	50–54	55–59	60–64	65–69	70–74	75–79	80–84
A	14.29	13.16	11.12	6.74	7.40	3.01	−11.25	−14.21
B	−15.66	−12.76	−3.00	−2.41	0.49	−6.43	−8.28	−9.38
C	−5.39	−4.48	−4.60	−4.23	0.46	−2.19	−6.19	−26.83

Source: Reproduced by the author with data from the Population Census by the Statistics Bureau, Ministry of Internal Affairs and Communications.

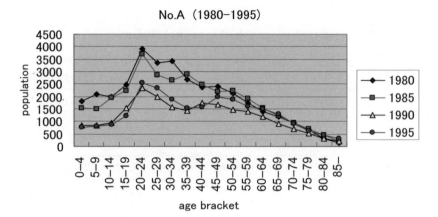

Figure 7.10 Population by age in District A (*Source*: Produced by the author using data from Population Census by the Statistics Bureau, Ministry of Internal Affairs and Communications)

since 1980 and is reflected in the inflow of young people staying near the level of the sustainable districts. Its demographic structure was comparable to that of District B in 1980, but is similar to that of a stable district circa 1995. Typically, the share of private rental houses has been increasing, with some variety in the floor area of dwellings. As compared with District B, which might be defined as a "new town" district, District C might be regarded as a "post-new town" district, in transition to a "stable

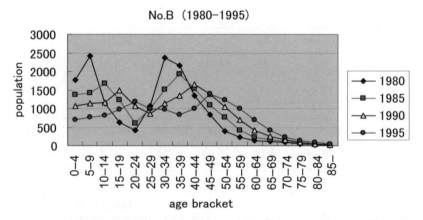

Figure 7.11 Population by age in District B (*Source*: Produced by the author using data from Population Census by the Statistics Bureau, Ministry of Internal Affairs and Communications)

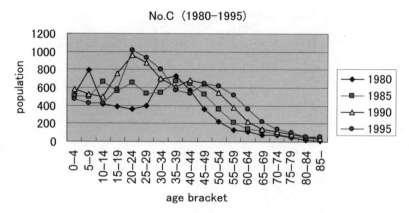

Figure 7.12 Population by age in District C (*Source*: Produced by the author using data from Population Census by the Statistics Bureau, Ministry of Internal Affairs and Communications)

Table 7.11 Composition of private households in control districts by ownership
1970   (%)

No.	Owner-occupied	Public rental	Private rental	Company houses	Lodgings
A	31.39	2.06	57.54	4.58	4.16
B	100.00	0.00	0.00	0.00	0.00
C	74.65	1.42	0.00	0.00	1.41

1995   (%)

No.	Owner-occupied	Public rental	Private rental	Company houses	Lodgings
A	38.33	1.63	50.59	4.62	1.79
B	16.61	76.61	5.37	1.09	0.19
C	36.57	11.92	43.84	5.92	0.40

Table 7.12 Composition of private households in control districts by floor area
1995   (%)

No.	Average floor area (m²)	0–29 m²	30–49 m²	50–69 m²	70–99 m²	100–149 m²	150 m²–
A	47.08	40.54	19.91	15.94	10.67	7.16	2.74
B	48.02	6.90	54.98	30.59	5.29	1.76	0.36
C	50.84	29.44	18.94	33.85	8.35	5.83	2.24

Source: Reproduced by the author with the data of Population Census by the Statistics Bureau, Ministry of Internal Affairs and Communications.

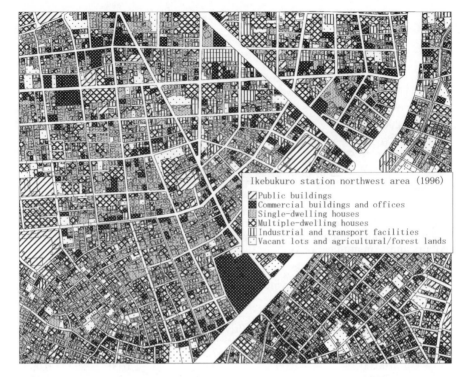

Figure 7.13 Land use map of District A (*Source*: Produced by the author using data from the Geographical Information System of the Urban Planning Bureau, Tokyo Metropolitan Government)

district''. The land use map in Figure 7.15 indicates a moderate mixture of different categories, which is typical of a stable district.

Thus, the afore-mentioned characteristics of the stable districts, that is the accommodation of young people and responsiveness to the diversity of residential status, become even clearer when compared with the control districts. At the same time, a range of characteristics can be observed between districts within the Tama New Town area.

5. Future prospects

Using population census data, this chapter considered the sustainability of a community in terms of population stability, with a special focus on residential function.

Our findings indicate that the stable districts have the common char-

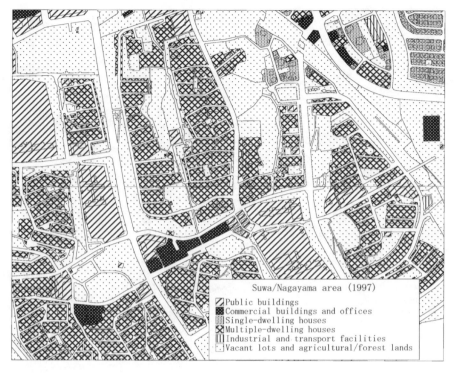

Figure 7.14 Land use map of District B (*Source*: Produced by the author using data from the Geographical Information System of the Urban Planning Bureau, Tokyo Metropolitan Government)

acteristics of responsiveness to the diversity of residential status and the existence of long-term residents who support the dynamism of the community. Some stable districts located to the west of the city centre are characterized by a constant inflow of young people. Supported by the existence of certain permanent residents, they also receive the benefit of revitalization by such young dwellers. Actors in such districts neither remain the same nor change drastically over a short period of time. Thus, those districts give the impression of flexible communities that are capable of accommodating progressive changes. Other stable districts correspond to single-dwelling residential areas characterized by the constant inflow of some family households. In this case, newcomer families gradually fit in with the local communities led by the established core actors of permanent residents. Those districts give us an image of communities where the primary local actors remain unchanged but constantly receive a certain level of stimulus. More detailed research will be necessary on

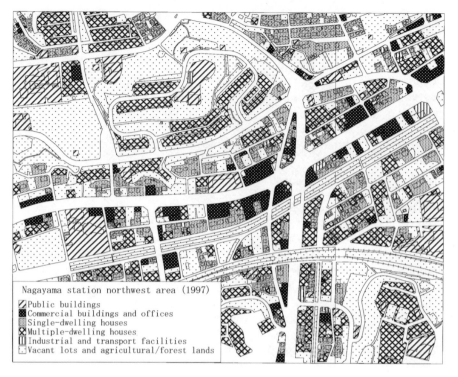

Figure 7.15 Land use map of District C (*Source*: Produced by the author using data from the Geographical Information System of the Urban Planning Bureau, Tokyo Metropolitan Government)

this type of stable district, including business accumulations in surrounding areas.

Populations returning to the heart of Tokyo have been observed in the current post-bubble period. The existence of those stable districts, however, suggests the possibility of living on the periphery, as opposed to the long-preferred option of living in a suburban single-dwelling house or the much-publicized option of living in the city centre. Also, the above-mentioned characteristics of the stable districts imply that the diversity of dwellings will ensure the accommodation of newcomers. It appears that areas with the proper levels of intensiveness and convenience tend to be preferred so long as housing prices stay within a reasonable range.

We should also look at some of the issues that remain to be addressed.

First of all, those stable districts happened to stay on a plateau throughout the 25-year survey period during the process of urban expansion. It is true that urban expansion in the Metropolis of Tokyo started with the restoration of imperial power in 1868 and continued into the bubble era. However, the premise of urban expansion may lose validity in the future, given that: (a) a third baby boom will not arrive due to the declining birth rate; (b) the geographical expansion of housing supply has come to a halt with the decline in land prices; (c) the share of convenience-oriented single-member households will continue to rise, and (d) stock accumulation has progressed to a considerable extent throughout the whole economy. In this context, the existence of stable districts should not be disregarded as a transitional phenomenon.

Attention should also be given to the qualitative difference in stability, according to the scale of urbanization. As already mentioned, revitalization by young dwellers seems to be instrumental in a large metropolitan area. In a smaller city, stability may require a concentration of functions including industries (workplace). There will, of course, be some levels in between. The existence of Type II districts implies the possibility of such stability.

Finally, the diversity of habitat preferences within a new town is another interesting trend, although it is treated as a secondary issue in this chapter. The Tama New Town has an inherent dual character. It exists as a district that has developed systematically through a new urban housing development project that coexists with an urban district developed through a land readjustment project. Districts B and C largely correspond to those two types of district respectively. If such trends observed in the two districts should generalize as a framework, future new towns may be divided into an intensive urban area and a nature-rich residential area.

A redevelopment project of a built-up urban area usually expands the floor area of buildings. Any technique of down-zoning that would transform a built-up urban area into a nature-rich residential area is not a practical option for now. If we are to ensure the "soft landing" of the generation-old new towns into a new phase, however, one of the options might be to redevelop them into nature-rich residential areas by way of contraction and refinement.

This chapter has provided a vision of local sustainability using case studies, and presented some hypotheses. As was mentioned at the beginning, the discussion presented here draws on a simple preliminary study. The definition of stable districts should be clarified by further case studies covering longer periods. Nonetheless, the finding that stable districts are not characterized by static stability but by perpetual changes that culminate in an integral framework should provide some implications for planning theory.

The stable districts might also be considered as typical examples of dynamic equilibrium, as described in Chapter 2. The only difference may be that land use should be mixed to some extent in order to retain the population, whereas some purification might be necessary so as to ensure the stability of land use itself, as indicated at the end of that chapter.

A critical mass of resources (including labour force and land) was required in the high economic growth period to sustain uniform reproduction on a progressive scale. In an era of stability and diversity, however, we would have to invest more wisely in limited resources. In this context, it would be unwise to adopt a development technique that further destroys nature in the raw in order to create nature-rich residential areas.

As in *Fragile dominion* by Levin (1999), "it may well be that natural systems are not so very fragile;... What is fragile, however, is the maintenance of the services on which humans depend". Are cities as a residential system, culturally made by human beings, fragile or not? This chapter reveals that the stability of population can be a key concept for the problem.

Note

The current land use/building status data used for analysis in this study was made available by the Planning Division, City Planning Bureau, Tokyo Metropolitan Government.

REFERENCES

Abe, T., Levin, S.A. and Higashi, M. (1997) *Biodiversity: an ecological perspective*, New York: Springer.

Burgess, E.W., Park, R.E. and Mckenzie, R.D. (1925) *The City*, Chicago: University of Chicago Press.

Common, M. (1995) "Sustainability", *Sustainability and Policy*, Cambridge: Cambridge University Press.

Davies, W.K.D. (1984) *Factorial ecology*, Aldershot, Hants: Gower.

Herbert, D.T. and Johnston, R.J. (eds) (1976) *Spatial Process and Form*, London: Wiley.

Jacobs, J. (2000) *The Nature of Economies*, New York: Modern Library.

Kawamura, K. and Hiroyuki, K. (1995) *Sustainable Community*, Gakugei Shuppan Sha (text in Japanese).

Levin, S. A. (1999) *Fragile Dominion: Complexity and the Commons*, Reading, Mass.: Perseus Books.

Meadows, D.H., Meadows, D.L., Randers, J. and Behrens III, W.W. (1972) *The Limits to Growth.* (A Report to The Club of Rome), New York: Universe Books.

Naito, M. and Saburo, K. (1998) "A Stable Social System", *Iwanami Lecture*

Course: Global Environmentology, 10, Iwanami Shoten, Publishers (text in Japanese).

National Land Agency (ed.) (1978) "The Third National Comprehensive Development Plan", Supplement to *People and National Land* (text in Japanese).

—— (1998) *The Grand Design for the 21ˢᵗ Century* (text in Japanese).

Oe, M. (1995) "Cohort analysis on changing population distribution in Japan: process of concentration on Tokyo area and future prospects", *Demographic Study* 51(3): 1–19 (text in Japanese).

Pielou, E. C. (1969) *An Introduction to Mathematical Ecology*, Wiley InterScience (http://www.3.interscience.wiley.com).

Shevky, E. and Bell, W. (1955) *Social Area Analysis: Theory, Illustrative Application and Computational Procedure*, Stanford CA: Stanford University Press.

Tamagawa, H. (2000) "A test study on the 'sustainability' of urban district", *Comprehensive Urban Study* 71: 5–20 (text in Japanese).

Tokyo Metropolitan Government (1998) *Draft Plan of Action for a Recycling-oriented Society* (text in Japanese).

8

Urban sustainability
A case study of environmental movements in Kamakura

Junko Ueno and Masahisa Sonobe

1. The sustainable city and public participation

1.1 The universalization of environmental problems and public participation

With the unprecedented devastation caused by the two world wars and global pollution on an epidemic scale, the twentieth century was definitely a "century of war and pollution" (Miyamoto 2000: 35). During the 1950s and 1960s, developed countries suffered from chronic severe pollution caused by industrialization and urbanization, the driving forces of mass production and mass consumption. Japan saw the emergence of a mercury poisoning disease called Minamata disease in Kumamoto Prefecture only ten years after the end of World War II. The four major "pollution diseases" (Minamata disease, Niigata-Minamata disease, itai-itai ("ouch-ouch") disease and Yokkaichi asthma) had already become major pollution problems by the 1960s.

Pollution and environmental destruction in industrialized countries led to protests by local communities. In Japan, public opinion and anti-pollution movements picked up steam in the late 1960s. The four major pollution lawsuits (filed between 1967 and 1969) and related support activities were monumental forces within Japan's post-war anti-pollution movements. However, those initiatives, which originated in support activities for the victims of industrial pollution, were by nature limited to action against private businesses. There appeared to be little chance that an institutional channel would be established to reflect the opinion of environmental or community groups in the process of environmental decision-making (Terada 1998).

Environmental issues in Japan reached a turning point in the mid-1980s. The period of "pollution and development" which lasted until the mid-1980s was followed by the period of "universalized environmental issues" (Funabashi 2001). Since the mid-1980s, attention has progressively shifted from the traditional industrial pollution problems to urban environmental problems related to daily life such as automobile exhausts, solid waste and domestic wastewater, as well as to global warming and other global environmental issues. The universalization of environmental issues has led to the diversification of environmental movements. Recycling movements have emerged in conjunction with global environmental issues, gaining support in the urban areas. Meanwhile, traditional nature conservation movements, which date back to the 1960s, have been gaining momentum along with new environmental movements, including national property and cityscape conservation.

The universalization of environmental problems has also transformed the relationship between environmental movements and environmental policy, as well as between citizens and government. The impact of environmental movements on environmental policy cannot be disregarded. Anti-pollution and anti-development movements that emerged in the 1960s, including the four major pollution lawsuits and the movements against the construction of petrochemical complexes in Numazu, Mishima and Shimizu (all in Shizuoka Prefecture), facilitated the promotion of pollution control measures in the 1970s, such as the adoption of 14 pollution control laws and the establishment of the Environment Agency. It was not until the 1980s and 1990s, however, that citizens were given a channel to express their opinion on environmental policy-making. With the acceleration of decentralization, citizens were increasingly encouraged to participate in the decision-making process throughout the 1990s.

The encouragement of public participation in policy-making and citizens' initiatives partly resulted from the fact that the government could no longer deal with environmental problems on its own. For example, collaboration between citizens, businesses and government is essential in making successful attempts to recycle. Mutual understanding is the key to ensuring effectiveness, whether citizens demand improvements on the part of businesses or government, or the government imposes unilaterally established criteria. Kamakura city, in its Basic Environmental Plan, has set the lofty target of reducing waste and CO_2 emissions by 20 per cent. "[Setting this target] would not be possible without public participation.

No bureaucrat would dare set this kind of target", said Ken Takeuchi, then Mayor of Kamakura.[1]

The responsibility of local governments in the field of the environment has expanded since the 1980s with the emergence of broad environmental issues. Local governments also have an important role to play in creating sustainable cities. Many of the activities intended to achieve a sustainable society, the main theme of the 1992 Earth Summit, are community-based, thus involving local governments as primary actors (Matsushita 2002). However, it is becoming increasingly difficult for local governments to address environmental issues by themselves. Collaboration with citizens is now required.

1.2 Significance of environmental movements

The importance of public participation is widely recognized today. The encouragement of public participation and citizen initiatives has become a national trend. For example, the Third Comprehensive Plan of Kamakura city endorsed public participation in its basic strategy by stating,

[A] mechanism for public participation in the policy-making process will be created to enable the active promotion of citizen involvement in municipal administration. It will also be mandatory to disclose data and information related to the administration and finance of the municipality and make them available as much as possible, so that every citizen may participate in municipal governance (Kamakura city 1993).

Previously, government would impose established policy on citizens and businesses in a unilateral fashion. Those days are over. We have now reached an era of collaboration between citizens, businesses and government. This applies not only to the environment, but also to a wide range of local issues including welfare and urban planning.

Environmental movements have also boosted the trend of public participation in the decision-making process. The number of direct requests for local referendums has been increasing, enabling local residents to express their opinion on development and environmental issues that concern them. Examples include the movement against the construction of the Ikego US Military Housing Complex in the City of Zushi in the 1980s, and the local referendum on the construction of a nuclear power plant in the town of Maki, Niigata Prefecture. It is true that few direct requests for local referendums have been accepted by the municipal assembly, showing the limited influence that environmental movements have. However, there is no denying the fact that environmental movements are oriented toward public involvement in the decision-making process. Movements related to environmental issues have worked as a non-institutional force that monitors and criticizes the decision-making process and social rationality of policy components (Hasegawa 2001).

This chapter focuses environmental movements in the context of public participation and specifically examines their collaborative linkages between citizens and government, taking the example of Kamakura city. The environmental movement is defined as "a collective action for change motivated by frustration over environmental conservation policy". Thus, environmental movements have three dimensions: discontent, orientation toward change and collective action (Hasegawa 2001). Frustration over environmental conservation policy does not mean personal anger, but discontent with environmental policy shared by many of the local residents. Such frustration tends to mount with the recognition of, for example, visible damage or the announcement of a development project. The term "orientation toward change" focuses on the objectives and values of the movements. Generally speaking, environmental movements are risk-averting in nature and seek concrete and direct solutions to the problems. Unlike personal protests, environmental movements have a collective and organized character.

Environmental movements are active in various fields. This chapter first provides an overview of environmental movements in Kamakura and their relations with the municipal government (Section 2), followed by a comparison of two contrasting environmental movements (Section 3). The two movements are based on the idea of common, not personal, ownership of land (environment). Focusing on this concept of common ownership of environment, Section 4 examines the "environmental" aspect at issue in Kamakura, which is being placed under common ownership. Finally, Section 5 presents an argument that environmental movements have the potential to develop civil society by encouraging public participation.

In today's society, marked by the universalization and deterioration of environmental problems, we need a social mechanism to promote environmental conservation, as well as technologies to solve the problems. Developing a community-based mechanism is an important task, but it is difficult for local governments to attempt this alone, so consultation with citizens is required. By describing environmental movements and discussing the idea of common ownership that underlies them, this chapter will demonstrate that local communities have finally become sufficiently empowered to decide on issues that concern them, after a long period of exclusion from the decision-making process.

2. Environmental movements in Kamakura

2.1 Development of environmental movements

"Any briefing session on condominium construction will turn into an anti-project rally", said one representative of a major real estate company (*Nihon Keizai Shimbun*, 30 September 1997, morning edition, p. 33). As part of the prestigious residential district Kamakura has experienced a continued decline in land prices in the wake of the economic bubble bursting. Newspaper articles attributed the slide in land prices to the development control. In order to preserve the value of the numerous historical sites, the municipal government of Kamakura has designated its old city area as a scenic district, where the law forbids unauthorized alteration of landscape. However, the remark of a developer quoted above implies that real estate companies have to cope with a deep-seated anti-development sentiment, in addition to the municipal government's meticulous regulations.

The citizens of Kamakura are known for their heightened awareness of the environment and their active pro-environmental conservation, anti-development stance. Having developed as a tourist resort since the Meiji era, Kamakura does not have a substantial indigenous population. A large part of the population consists of those who moved in after World War II and their families. Suburban development in Kamakura accelerated in the rapid economic growth period, as the population increased sharply from 98,617 in 1960 to 165,552 in 1975 (167,583 in 2000). Many of those newcomers were fascinated with the environment of Kamakura, which seems to be one of the reasons why Kamakura residents have a keen interest in the environment.

The number of environmental groups attests to the degree of their interest in the environment. Table 8.1 shows the number of environmental groups in the major cities of the Kanagawa Prefecture and the number of environmental groups per 10,000 residents.[2] In terms of the absolute number of groups, Yokohama leads with 141, second-ranked Kamakura has only 24. However, the number on a population basis shows a different picture, as Kamakura is ranked first in Kanagawa Prefecture with 1.43 environmental groups per 10,000 residents.

The unique character of Kamakura is not solely the keen interest of its residents in the environment, but its links with civic movements and activities. There are 236 groups registered with the Kamakura NPO Support Centre (April 2002). Many groups have multiple objectives. Seventy-six of the registered groups include "environment/conservation" in their objectives and are involved in various activities such as the pro-

Table 8.1 Comparisons between population figures and environmental group numbers

City	Population	Number of groups	Number of groups per 10,000 residents
Kamakura	167,254	24	1.43
Zushi	57,112	6	1.05
Minami-Ashigara	44,221	3	0.68
Hiratsuka	254,000	16	0.63
Miura	52,964	3	0.57
Yokohama	3,373,110	141	0.42
Kawasaki	1,231,851	14	0.11

Source: Kanagawa Prefecture statistics (March 1999).

motion of recycling, cleaning, raising environmental conservation fund and environmental learning.[3]

The vitality of civic movements and activities on the environment in Kamakura can be partly explained by its historical background. The Kamakura Trust, arguably the first national trust movement in Japan, was born in this city (Kihara 1998). Thus, the city has a long history of anti-development and environmental conservation movements. Environmental movements in Kamakura date back to the pre-war period. Kamakura Dojin-Kai (a group of like-minded people), established in 1915, conducted such activities as the protection of cedar trees along Wakamiya Oji Street at the Tsurugaoka Hachiman Shrine, repair of Dankazura (pedestrian path in the centre of Wakamiya Oji Street) and reconstruction of the station building, with the objective of "protecting historical materials and sites, increasing access to education and sanitation, improving public morals, promoting industries, and conducting other activities beneficial to the public (Sawa 1965). Its membership included influential politicians, businessmen and military officers who spent their money on the protection of Kamakura's historical environment in order to improve residents' livelihoods.

Civic action in Kamakura since the pre-war period may be classified into three periods with different actors and relationships with the government (Fujii 2000). The first generation of civic activists includes influential politicians, businessmen and military officers who owned country houses there in the Meiji and Taisho periods (late nineteenth and early twentieth centuries). Kamakura Dojin-Kai is typical of civic action in this period. The group's activities complemented government activities and tasks were not duplicated. The post-war generation was comprised predominantly of intellectuals who have promoted innovative landscape conservation and traffic control measures. One typical example is the movement against the development of Oyatsu (a hill behind Tsurugaoka Hachiman Shrine) led by Jiro Osaragi, a popular writer. This movement led to the establishment of the Kamakura Trust, the first national trust in Japan (1964), and the enactment of the Law Concerning Special Measures for the Preservation of Historic Natural Features in Ancient Cities (1966). Although ideas and movement in this period subsequently brought about a reformist municipal administration, severe antagonism between civic movements and the municipal government ensued on the return of a conservative mayor. The third generation of civic movements emerged in the 1990s, this time led by average citizens, particularly women. Both citizens and the municipal government matured as a result of the confrontation, which paved the way for coordination and collaboration. The result was the development, in 1996, of the Third Comprehensive Plan, which advocates the creation of an environment-friendly community that entails the active involvement and action of the citizens. The establishment, in 1998, of the Kamakura NPO Support Centre, sponsored by the municipal government, clearly demonstrates that civic action has taken root. The change in actors and relationships with the government also applies to civic movements and activities related to the environment.

2.2 Trends in public participation and the Kamakura City Comprehensive Plan

Since the end of World War II, the relationship between citizens and the municipal government of Kamakura has largely changed from that of confrontation to one of partnership. Apparently, this relationship has been influenced considerably by the policy of the mayor.

This subsection takes the example of the Kamakura Citizens Round Table Conference on Urban Planning (renamed Kamakura Citizens Round Table Conference on Community Building), held in 1985, to present a brief history of interaction between the mayor and public participation (Figure 8.1). The Kamakura Citizens Round Table Conference on Urban Planning (Round Table Conference) was established with some 200 members in 1972, in the wake of civic movements against the construction of condominiums in Komachi and Yukinoshita districts. Its parent bodies included the Association for the Conserving the Nature in Kamakura and the Network for the Master Plan of Kamakura (Kamakura Citizens Round Table Conference on Community Building 1999). With seven working groups, the Round Table Conference had some bearings on the adoption of a Citizens' Charter, the preparation of basic strategy and zoning regulations.

The establishment of the Round Table Conference was facilitated by

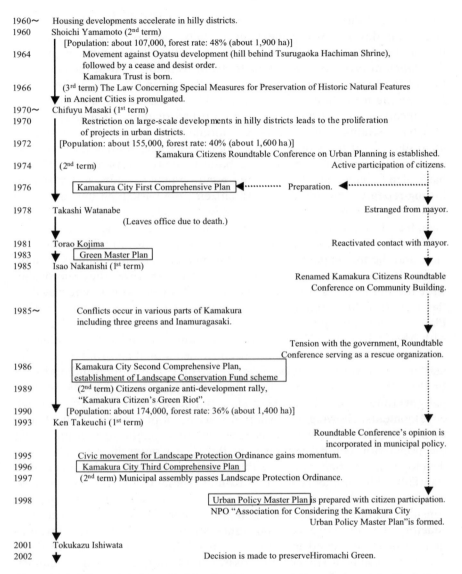

Figure 8.1 Mayoralty, Roundtable Conference and Development in Kamakura (*Source*: produced by the author)

the election of Chifuyu Masaki as Mayor, backed by the Kamakura Citizen's Union, a coalition force of the Socialist and Communist Parties with labour unions. The election of a reformist mayor, who had committed himself in favour of public participation, substantially facilitated the involvement of citizens in decision-making and led to the deterrence of large-scale development projects in the hilly districts. However, the control on developments in the hilly districts resulted in the concentration of development projects in urban districts. Movements against the construction of condominiums in various parts of the city in the early 1970s culminated in the formation of the Round Table Conference as a forum to ensure incorporation of public opinion in urban planning. The preparation of a basic strategy for the First Comprehensive Plan must have been a huge success for the Round Table Conference. The Kamakura City First Comprehensive Plan consists of a basic strategy (prepared in 1976), a basic plan (decided in 1979) and an execution plan (decided in 1980). Following repeated consultations with citizens, the basic plan unequivocally calls for the protection and preservation of historical and natural heritage, based on public participation in community building.

The initially friendly relationship between the Round Table Conference and the municipal government became erratic with changes in mayoralty. When a conservative mayor took office in 1978, the relationship with the municipal government became estranged with the removal of the Round Table Conference representative from the Kamakura City Planning and Zoning Commission. Following the reactivation of contact with Mayor Kojima, elected in 1981, tension returned with the election of Mayor Nakanishi in 1985, followed by occasional confrontations (Kamakura Citizens Round Table Conference on Community Building 1999). In 1985, the draft Second Comprehensive Plan, prepared by the municipal government, was to accommodate population increases and housing developments, allowing a substantial increase of population from 175,000 to 240,000 (Komai 1985). This represented a major turnaround from the population control measures in the First Comprehensive Plan to a development-tolerant policy. Public participation was rolled back substantially. The influence of the Round Table Conference on policy-making was reduced to simple provision of advisory opinion, and the consultation between the municipal government and citizens was not satisfactory. Under the Nakanishi administration, conflicts with development projects occurred in various parts of the city, including the three major green hills of Hiromachi, Tokiwa and Daimine, as well as Inamuragasaki Cape and the hill at the back of Engakuji Temple. When Mayor Nakanishi adopted a policy to "authorize development in accordance with the law", activist citizens began to appeal their complaints to the Round Table Conference and the Round Table Conference became the last fort of environmental movements (Kamakura Citizens Round Table Conference on Community Building 1999).

The situation changed again in 1993, when Ken Takeuchi was elected mayor. Following the election of an environment-conscious mayor supported by a citizens' group, the opinion of the Round Table Conference was reflected in municipal policies to a considerable extent. In 1995, the Third Comprehensive Plan was finalized with the aim of establishing citizen autonomy and building a humanitarian and environment-friendly community. Public participation was finally institutionalized in 1996 with the adoption of a Kamakura City Urban Policy Master Plan. The Association for Considering the Kamakura City Urban Policy Master Plan, a non-profit organization, was established to promote the implementation of the Master Plan.

Thus, public participation in Kamakura has experienced ups and downs depending on the policy of the mayor. Accordingly, the development policy oscillated between tolerance and restriction. Citizen opinion has now come to be reflected in municipal policy, as attested by the acceptance of the long-time demand of citizens for the preservation of Hiromachi Green. Environmental movements continued actively even in the period of estranged and tense relationship with the municipal government. Finally the government recognized citizens as non-negligible actors, resulting in the present policy of promoting citizen autonomy and environmental conservation.

3. Two environmental movements

3.1 Anti-development movement in a hamlet

Civic movements in Kamakura have exerted considerable influence on the municipal government, serving as the driving force behind the establishment of the Kamakura Trust, the promulgation of the Law Concerning Special Measures for Preservation of Historic Natural Features in Ancient Cities and the formation of the Kamakura Citizens' Round Table Conference on Urban Planning. Most of the civic movements described in this chapter were born out of frustration over environmental conservation policy, and hence should be regarded as environmental movements.

Environmental movements in Kamakura may be classified into anti-development or environmental conservation. Anti-development movements primarily aim at preventing specific development projects, including the construction of condominiums in Komachi and Yukinoshita, which directly resulted in the formation of the Round Table Conference. Environmental conservation movements advocate landscape conservation such as the preservation of Hiromachi, Tokiwa and Daimine Greens. Most of the environmental conservation movements originated from anti-development activities, including the case of Oyatsu. However, they are

primarily interested in landscape conservation after blocking developments, rather than the fight against development itself. For the purpose
of this chapter, environmental conservation movements are distinguished
from anti-development movements according to the primary objective,
although the two are closely related to each other.

What is the difference between an anti-development movement and an
environmental conservation movement, apart from their primary objectives? This section seeks to answer this question by taking the examples
of an anti-development movement in a hamlet and a landscape conservation movement for the three major greens in Kamakura.

Just around the time when the Oyatsu development project gave rise
to controversy, an anti-development movement emerged in Hamlet S
(see Figures 8.2 and 8.3), located in a valley at the eastern end of Kamakura. This movement was one of the first of its kind in Kamakura after
the war. Since then, three protests have been filed in Hamlet S, all resolved in favour of the residents. The first phase of the movement concerned a housing development by the Seibu Construction company,
which started in 1960. The protest started in 1966 and ended with an
agreement in 1971. The second phase, concerning the basic landscape
plan of the municipal government, continued from 1990 to 1996. The
third phase of the movement, in 1995 and 1996, was triggered by housing
construction for Seibu Real Estate on the site targeted by the previous
two protests. Who were the main actors of the movement in Hamlet S?
How did it develop? A detailed description of the three protests follows.

The first protest was filed against a housing development by Seibu
Construction (see Chronological Tables: Table 8.2). In the 1960s, the
whole area of Hamlet S suffered serious damage from flash floods and
landslides. They were caused by the construction of a golf course in 1960
and 1961, which left an extensive hill area deforested. Heavy rains caused
water to flow from the construction site all the way to the farthest end of
the hamlet, sometimes carrying logs used in landslide prevention. A large
rock even tumbled down to the house at the rear end of the hamlet,
adjacent to the housing construction site. Subsequently, land ownership
of the hill was covertly transferred to the Seibu Railway Company, and
Seibu Construction started housing construction works as contractor.
The hamlet continued to suffer damage. It was recommended by a city
councillor, who witnessed the extent of the damage, that a movement be
launched in defence of Hamlet S. In 1966, a petition for effective measures against flash floods caused by the housing development was submitted to the city council, with the signature of some 100 households. Despite these efforts, the municipal government failed to take any effective
measures, and the Association in Defence of Hamlet S was formed in
November 1969. The Association's membership was largely made up of

Figure 8.2 Map of Hamlet S and surrounding area (*Notes*: Hamlet S is the housing area at the centre of the map. Aligned on the lower right is the residential area developed by Seibu Construction. *Source*: Map provided by Increment P Corp., 2002)

households that had actually suffered damage. The Association had no rules such as by-laws and procedures for electing the chairperson and representatives. An environmentally-minded city councillor joined the Association as a supporter. The Association in Defence of Hamlet S was a loose-knit group "formed only because collective bargaining with Seibu and the municipal government is more advantageous for us than individual negotiations by the residents concerned" (interview with a hamlet representative, 31 August 2002). The group did not strictly define its membership partly out of consideration for the landowner's relatives

Figure 8.3 Hamlet S, view from the green belt, 15 September 2002 (*Source*: Photo by authors)

who lived in the hamlet. Since some landowners were in favour of, and therefore had an interest in, the housing development, residents were not necessarily united in the protest against Seibu Construction. It is for this reason that a voluntary association of residents was formed to collect signatures. On the introduction of the supporting city councillor, one of the hamlet representatives made contact with those active in the Oyatsu movement for advice on how to proceed with the protest, including the collection of signatures and the filing of petitions.

The problem was resolved by an agreement in 1971, when Seibu Construction offered to conserve greenery and construct a public park. In compensation for the loss of sunlight and landscape, Seibu Construction created a green belt to serve as a buffer zone between the housing development area and the hamlet. The company also constructed a public park in the hamlet for use by local residents (which was subsequently donated by the residents to Kamakura city). Thus, the movement came to an end when the residents' demands were accepted.

The second wave of protest was directed against the municipal government's landscape improvement plan. In 1990, the municipal government of Kamakura issued a statement on the basic landscape plan for the green belt created in the first phase of the movement (a buffer zone between the hamlet and the housing construction area on the hill). Local residents happened to find out about the plan when a man broke through the green belt and threw stones at a house at the rear end of the hamlet. A city official who came to verify the damage informed the residents about the plan. A petition was filed to protect the security and privacy of the residents, out of concerns that the plan might be implemented without notification, and that the landscape improvement might increase the risk of similar damage. The petition, filed in the names of the two leaders of the neighbourhood groups adjacent to the green belt, called for changes to the nature trail routes and installation of fences, so as to make sure that no park would be created and entry would not be allowed to the green belt. In 1991, the then-Mayor Nakanishi sent a summary of the landscape design improvements to the leaders of the neighbourhood associations concerned. The document stated that the residents would be informed of details on the landscape improvement works, which was planned to start in the fiscal year of 1992.[4] Subsequently, however, some of the works, including tree cutting, were implemented without any explanation to hamlet residents. Following the submission in 1994 of another petition to the Construction Department of the municipal government, a city councillor coordinated a meeting between the Green Division of the municipal government and six volunteer residents, including Ms Eda (pseudonym). Although Ms Eda, who lived at the opposite end of the hamlet, did not have any direct interest in this matter, she was invited to participate in the protest for her experience as principal actor in the first stage of the movement (interview with Ms Eda, 31 August 2002). The second phase of the movement came to an end in 1996, when the Municipal Park and Landscape Division revised the plan, which incorporated residents' demands, including the installation of facilities to prevent access from the green belt to the hamlet and the modification of nature trail routes.

The third phase of the movement overlapped in time with the final stage of the second period. It began in 1995, when residents filed a protest against Seibu Construction, claiming that the two houses for installment sale, constructed on the edge of the green belt, violated the agreement concluded in 1971. Negotiation between local residents and Seibu started with a review of the 1971 Memorandum of Agreement. It soon became clear that the local office director of Seibu Construction did not know about the memorandum, and that the municipal government had authorized the project on the grounds that it did not violate the law. The residents protested against the sale of the new houses saying, "our privacy will be jeopardized as we shall be in plain view of the prospective inhabitants of the houses that have suddenly sprung up on the green hill behind us". Supported by two city councillors, representatives of the affected households, the residents opened negotiations with Seibu. In 1996, they accepted a proposal by Seibu to plant tall trees in front of the new houses to block the view. As Seibu Construction responded swiftly to the demand of local residents by presenting relevant drawings and offering an apology, the third phase of the movement was concluded in just six months. The initial agreement between Seibu and local residents envisaged planting eight trees, one of which was subsequently removed at the request of a purchaser.

In comparison with the first phase, the second and third phases of the movement did not involve collective action by an overwhelming majority of hamlet residents. The municipal government and Seibu both responded swiftly to the demands of residents. Indeed, access to drawings related to the development plan, which was difficult during the first phase, was provided in a short period of time in the second and third phases. Although the three phases differ in length of period as well as in terms of the existence or non-existence of a core institution and signature collection, the movement had these three characteristics in common: (1) concentration on a small area, (2) leadership by those who suffered actual damage, and (3) termination with the solution of problems.

The movement was confined to a small area, whether it involved the whole Hamlet S or some of the groups in the hamlet (1). The neighbourhood association that includes Hamlet S (comprising some 1,700 households) was only involved in the movement when the mayor responded to the association in the second phase. Otherwise, the movement concerned some 150 households in Hamlet S, and had little interaction with external actors. In the first phase, Ms Eda was supported and guided by one of the primary actors in the Oyatsu movement, and the Association for the Protection of Greens and Historic Sites in Kamakura. However, this was because Ms Eda was Vice-Chairman of the Association and many of the members lived in the hamlet. In fact, other members of the Association in Defence of Hamlet S did not know about this support. Although individual members were invited to meetings of other civic groups in Kamakura after the end of the first phase, the Association in Defence of Hamlet S conducted no coordinated activities with other groups either during or after the protest. The relationships between individual members of the Association as well as between them and other groups were never particularly close, but were based on loose personal contacts.

The geographical limitation of the movement may be attributed to the fact that the major actors were those who had suffered actual damage (2). It is for this reason that the first phase of the movement involved the collection of signatures throughout the entire hamlet. During the second and third phases, however, the primary actor of the movement was a group of residents who lived at the rear end of the hamlet, near the green belt. The main objective in each phase was to redress actual damage, and the movement ended in an agreement with the other party to put an end to the damage or to prevent further damage (3). The movement has never been oriented towards an absolute ban on development or active conservation of the green belt after saving it from development.

3.2 Conservation movement for three major greens

This subsection describes the movement for the conservation of three major greens in Kamakura (Hiromachi, Tokiwayama and Daimine), which is in stark contrast to the anti-development movement in Hamlet S.

In Kamakura, various civic groups are involved in the landscape conservation movement for the three major greens of Hiromachi, Tokiwayama and Daimine. Since the Oyatsu movement in the 1960s, the issue of landscape conservation has driven many people to collective efforts. The collection of tens of thousands of signatures for conservation of these greens attests to the scale of mobilization on this issue.

It was in 1973 when a housing development project emerged in Hiromachi, one of the three major greens (see Chronological Tables: Table 8.2). Three developers announced a draft project, but the procedure was soon suspended in the face of protests by local residents. In 1983, the three developers submitted a new application to the municipal government for preliminary review of the development project. Following the presentation of 60,000 signatures protesting against the project, the municipal government put the development procedure on hold. In 1984, resident and neighbourhood associations in the vicinity of the green formed a Joint Association for the Protection of Kamakura's Natural Environment. The Joint Association submitted 120,000 signatures to the prefectural government of Kanagawa in 1990, calling for the designation of Hiromachi as a landscape conservation area. In 1995, 220,000 signatures were presented to the City Council of Kamakura for the enactment of a Kamakura City Landscape Conservation Ordinance. Meanwhile, Ken Takeuchi, the then-Mayor of Kamakura, announced a revision to the basic policy for landscape conservation and refused to accept the application for preliminary review submitted by the developers (1993). In 1996, the municipal government of Kamakura rejected the application for development permission. The developers filed a request for reconsideration with the prefectural Development Review Board, which, in 1997, adjudicated the rejection by the municipal government to be null and void. Kamakura city created a Green Conservation Fund scheme in 1986 and entered into consultation with the prefectural government for possible conservation measures, setting aside ¥11.4 billion by 1998. The Joint Association for the Protection of Kamakura's Nature, for its part, launched a trust movement in 1998, led by participating resident associations. Donations rose to ¥24 million in just three months. With the addition of the Shichirigahama Neighbourhood Association in 2000, the membership of the Joint Association includes some 4,500 households in the following nine resident and neighbourhood associations: Kamakurayama Neighbourhood Association and Kamakurayama-Wakamatsu Residents Association in the northeast of Hiromachi; Minami-Kamakura, Shin-Kamakurayama and Goshogaoka Residents Associations on the west of Hiromachi; and Shichirigahama Residents Association, Shichirigahama-Nichome Residents Association and Shichirigahama Neighbourhood Association in the southern coastal area. By October 2002, the Joint Association had collected ¥33,542,766 in 3,358 donations for the Kamakura-Hiromachi Green Trust.

After 20 years of conservation movement, Hiromachi Green is now expected to be conserved in its entirety, as the municipal government of Kamakura reached a basic agreement with the three developers in October 2002 on the purchase of the 36.4 ha area reserved for the development project.

Similarly, local residents have been orchestrating a tenacious movement to conserve Tokiwayama and Daimine Greens. An extensive collection of signatures has been accompanied by fund-raising for the conservation of these greens. In 1990, 120,000 signatures calling for the designation of Tokiwayama as a landscape conservation area were handed over to the prefectural government of Kanagawa. Kamakura city has been purchasing Tokiwayama (about 20 ha) for almost a decade. The total area purchased from the Mitsubishi Estate Company has almost reached one-third of the green hill.

The conservation movement for Daimine was triggered in May 1976, when the Nomura Real Estate Development Company. announced a plan for a housing development in the area. The original development plan of Nomura Real Estate, tentatively named "Nomura Kamakura-Daimine Housing Site" (total development area: about 27 ha; planned number of houses: 80) was changed in 1996 to a land readjustment project involving many individual landowners as well as Nomura, tentatively called "Yamazakidai Land Readjustment" (total development area: about 28.8 ha; planned number of houses: 650). Answering the call of the "Kitakamakura-Daimine Trust: Fund for Conserving Landscape in Kitakamakura for Future Generations", over ¥10 million (¥10,506,110) was donated to the landscape conservation fund by 2002. Kamakura city also plays an active role in landscape conservation, seeking to preserve Daimine Green by extending the Kamakura Central Park, located in its vicinity.

In comparison to the anti-development movement in Hamlet S, the conservation movement for the three major greens is characterized by: (1) wide geographical coverage, (2) unimportance of actual damage in taking part in the movement, and (3) the potential for sustained activities.

The scale of signatures and donations for the three major greens indicates that the conservation movement has reached the whole city of Kamakura, or even beyond the municipal boundary (1). Although local communities are the main actors in these movements, support comes from a wide range of people including those who are not directly affected by the development (2). Tens of thousands of signatures collected imply that most of the people are not in daily contact with the green hills. The objective of the movement is to conserve the greens for future generations (3). Thus, the movement is not limited to spontaneous protests to prevent development but also considers how to conserve landscape and how to ensure the on-going maintenance and management of the greens. "We will make proposals based on the citizens' vision of municipal forests", said Shohachi Oki, Director-General of the Joint Association for the Protection of Kamakura's Nature (Kanagawa Shimbun, 4 October 2002). His remark expresses the intention to seek active involvement in the decision-making process on landscape conservation and to follow up on government decisions.

The comparison between the anti-development movement in a valley hamlet and the environmental conservation movement for three major greens reveals differences in: (1) the scope of the activists and supporters, (2) the importance of actual damage as motivation, and (3) the durability of action. Those two types of environmental movement are diametrically opposed in many ways, setting aside their differences in objective (protesting against a specific development or environmental conservation).

Nonetheless, they are not completely different types of movement. Indeed, the two types of environmental movement have one thing in common: the concept of common ownership. The following section examines this concept.

4. Common ownership of the environment

4.1 Basis for environmental movements: concept of common ownership of environment

The anti-development movement in Hamlet S and the conservation movement for the three major greens both question the use of land owned by others. Why can the actors protest against a development or call for landscape conservation on land owned by others? Why do the government and developers accept such demands? This subsection looks into the ground for those demands.

The two environmental movements may be considered to challenge private land ownership. Urban land use problems underlie environmental movements. Claims on such land use provide the basis for the assertion of rights, which is expressed in the form of a demand for public participation (Nitagai 1976: 354). In the first phase of the movement in Hamlet S, life-threatening damage from development activities literally made it a matter of life and death for local residents forcing them to demand the right to live in peace and security. In comparison, the second and third phases of the movement in Hamlet S and the conservation movement for the three major greens did not involve any life-threatening danger for local communities. Still, those protests were also led by local residents who asserted their right to enjoy the peace and amenities of their home town. Thus, the environmental movements are an expression of protest against the urban land use system that gives priority to private ownership, based on the very fact that the residents "live" there.

The environmental right, as first advocated in 1970, was the right of residents who "live" in the area. "Recommendations for the Establishment of Environmental Right", issued in 1971 by the Study Group on Environmental Right of the Osaka Bar Association, states that the environment is a property essential for the survival of human beings, and hence is to be co-owned by all people, and that any damage to the environment without the consent of its co-owners is illegal, whether or not it inflicts any damage on human health and/or their living environment (Study Group on Environmental Right, Osaka Bar Association 1971). By regarding the environment as the common property of all people, the Recommendations seek to place a limit on private property rights.

Environmental right is a concept that ensures "participation" of local communities in the planning of projects involving substantial modification of the environment, or argues for the necessity of obtaining "agreement" from local residents in modifying the environment. It attaches more weight on the "fact" that the environment is shared than on individual rights (Seki 2001). Thus, environmental rights are not given to individuals, but to groups of people who share the same environment.

In this context, it can be said that the Japanese tradition of "collective ownership" underlies the environmental rights, as advocated in 1970, and subsequent environmental movements. "Collective ownership" is a property right of a community (village) and the source of private ownership. Some villages effectively limit private property rights on legally owned land, so that individuals may not have complete freedom to dispose of their land or change farmland into housing lots. In Ryujin Village in Wakayama Prefecture, for example, a villager who moves out usually sells his/her residence and forest land to another villager, out of the consideration that "village land should be kept in the hands of villagers whenever possible". In this way, the freedom of transaction is restricted on a non-legal basis (Fujimura 2001: 40). Village land is everyone's "common property", and the right of individuals to dispose of their land freely is limited by the "collective ownership" of the village. A common is the most typical expression of such restriction.

Substituting the term "collective ownership" with "shared right of possession", Hiroyuki Torigoe notes that the concept of "shared right of possession" is a typically Japanese tradition, which still prevails in cities as well as in rural communities (Torigoe 1997). That anti-development and environmental conservation movements gain ground and are recognized by government and developers attests to the existence and functioning of this shared right of possession. According to Torigoe, the shared right of possession refers to the fact that local residents collectively take possession of a specific area on the grounds that they effectively "utilize" the land by simply "living" there (Torigoe 1997: 95). Thus, the fact of "living" justifies the claim of possession, effectively giving a group of, and not individual, residents collective bargaining power against landowners and developers.

4.2 Definition of "environment"

Environmental movements are based on the idea that local communities have the shared rights of possessing the environment. What kind of environment, then, do local residents share? Any difference in the idea of the environment to be shared would affect judgement on what kind of "environment" should be protected.

The term "uniqueness of Kamakura" is used repeatedly in environmental movements in the city. A book edited by the Kamakura Citizens Round Table Conference on Community Building begins with the following passage on the "uniqueness of Kamakura."

Every discussion and petition (on environmental conservation) cites the "uniqueness of Kamakura". The meaning of "uniqueness of Kamakura" as an abstract term is fully understood by all parties involved. Elements of this "uniqueness" come to everyone's mind, including the city's history since the Kamakura Shogunate, cultural heritage, distinct topographical features (mountains and the sea) and beautiful nature-rich landscape (Kamakura Citizens Round Table Conference on Community Building (ed.) 1999).

For citizens involved in environmental movements, the vague and abstract expression "uniqueness of Kamakura" represents the environment to be protected. No doubt, every citizen has a different definition of "uniqueness". It is for this reason that discussions often boil down to the meaning of "uniqueness of Kamakura", i.e. the definition of the environment to be protected.

One such discussion occurred during a confrontation between the citizens and the municipal government over the construction of condominiums. In Kamakura, condominium construction projects in the urban area have often triggered protests. In October 1997, resident associations near seven condominium construction sites in the city formed a Joint Association against Condominium Construction in the Ancient City of Kamakura, afraid that condominium construction would spoil the landscape of housing areas in the ancient city (Environmental Protection Association "Appeal from Kamakura" 1997). However, the conflict between condominium construction and cityscape conservation was virtually excluded from the scope of "environmental issues" as defined by the municipal government (Nakazawa et al. 2001). The Third Comprehensive Plan of the Kamakura City aims at developing attractive living environment along with admitting various types of housing, effectively pointing to a policy of developing urban areas while at the same time conserving green hills. For the municipal government, the construction of condominiums in the urban area helps reverse the decline in population, far from destroying the "uniqueness of Kamakura", as claimed by the protest movement.

Controversy also emerges on the value of environment. In discussing what kind of environment should be protected, opinion is sometimes divided over the value of the environment: some emphasize academic value while others underline historical value, for example. As regards the decision to fully conserve Hiromachi Green in October 2002, the Joint Association for the Protection of Kamakura's Nature appreciates the environmental impact assessment by the prefectural government, saying that Hiromachi obtained the highest rating in terms of ecosystem and forest landscape. However, some question the values of Hiromachi Green, suggesting that it does not have any historical significance unlike Oyatsu, nor does it have any ecological significance because Hiromachi is isolated from other green hills. Paradoxically, even the Joint Association overlooks the fact that Hiromachi is a treasure house of historical monuments from ancient to modern times and still retains its original form as a medieval mountain castle (Yamamoto 1997).

Also, local residents may have different visions of what constitutes "environment". In the third phase of the anti-development movement in Hamlet S, old-time residents demanded that tall trees be planted in front of the new houses on the hill to block the view of the hamlet. Subsequently, however, one of the trees was removed as a new resident protested against the tree planting, saying: "I paid for a house with a view" (interview with an old-time resident in Hamlet S, 15 September 2002). To old-time residents, the newly constructed houses on the hill represent nothing but a nuisance. It is unbearable to see such a deformation of the traditional landscape. As for the purchasers of the houses, however, they accept the local landscape including the new houses constructed by the developer. They would never have thought that some people might regard their houses as unexpected intruders. Thus, old and new residents may well have different understandings of the "uniqueness of Kamakura". This divergence in views might become another cause of conflict.

The final analysis shows that the vague notion of "environment" allows people to have different definitions of what should be protected. As far as the green conservation movement is concerned, not only local residents but also tourists and intellectuals are mobilized to disseminate information through the media and stimulate discussions. In this context, a green hill transcends a mere physical presence and enjoys a symbolic meaning (Nakazawa et al. 2001). It is true that this symbolic, rather than physical, significance of "environment" as something to be upheld may lead to disputes over its interpretation. However, it also enables a larger number of people to join the movement, as we have seen in the case of the green protection movement.

5. Will a civil society be built?

This chapter has examined the idea that underlies two contrasting environmental movements: the anti-development movement and the environmental conservation movement. In terms of scope and viability, the anti-development movement is considered unlikely to promote public participation in Kamakura. The movement, being confined to a very small area and ending with the implementation of damage control measures, has little motivation toward sustained involvement in land use decision-making for the entire city of Kamakura.

However, the idea that underlies the two movements gives us the impression that the anti-development movement and the environmental conservation movement both have the potential to stimulate public participation in municipal governance, thus paving the way toward the development of a civil society. The two movements find their justification in the concept of common ownership, which grants the right of possession based on the simple fact that local residents "live" in the area. The concept of common ownership entitles local residents to negotiate with landowners, government and developers on an equal footing and grants the right of decision to their collective will. The concession made by developers and government also attests to the acceptance of this concept in Japanese society.

Environmental movements serve to remind people of the concept of common ownership. The expansion of environmental movements makes people realize that they have a say in community matters and encourages active involvement in decision-making.

In some environmental movements, conflict of opinion does arise over what kind of environment should be protected. That the environment means different things to individual activists may give rise to confusion in action or in policy-making. However, the same diversity in approach enables the mobilization of various actors. What is important is to open discussion to various points of view and not to exclude different views in the search for consensus.

In a world where environmental problems are worsening on a global scale, technical solutions to environmental problems are insufficient to ensure the viability of urban society. A social mechanism is needed to counteract environmental problems or to promote environmental conservation. As seen in the example of the four major pollution diseases, government and businesses sometimes undermine the security and environment of local communities. The establishment of sustainable urban society requires a mechanism that allows citizens to monitor the activities of government and businesses and promote environmental conservation. However, it has been difficult to date for citizens to secure a level playing field as the interest of government and businesses has, in most cases, prevailed over their rights. The idea underlying environmental movements clearly indicates that the collective right of local residents need never succumb to the interest of government and businesses, and mobilizes citizens for active involvement in community issues. That is why this chapter focuses on environmental movements. If we are to build a civil society,

Table 8.2 Chronological Tables

1. Movement in Hamlet S against a housing development project by Seibu Construction, 1960–1971

1960   D: Construction company levels the land for constructing a golf course.
1961   • Heavy rainfall causes hamlet roads to be covered with water. Mudslides cause damage (which will continue into 1965).
1963   D: Seibu acquires land for housing construction.
1966   D: Seibu is granted development permission. Seibu Construction starts work.
        • Heavy rainfall causes mudflows on roads, carrying logs and stones.
        R: A Kamakura City Councillor visits the site in response to a phone call from a hamlet resident.
        R: A petition for damage control measures is submitted to Kamakura City Council (signed by about 100 households).
1967   D: First briefing by Seibu Construction (for local residents and the municipal Construction Department).
        D: Housing land development makes headway, with the construction of a 30 m-high barrier on the cliff at the back of the hamlet.
1968   R: Negotiation starts with Seibu Construction.
        D: Seibu starts and completes sewage repair works in the hamlet.
        • The barrier under construction partially collapses due to heavy rain. Muddy water flows into houses at the bottom of the barrier.
        B: Seibu Construction and resident representatives exchange memorandum of understanding on the compensation of damages caused by the construction (in the presence of the City Councillor).
1969   R: Hamlet residents report damages caused by the development project in a symposium on "Land Development and Landscape Conservation" organized by Yomiuri Shimbun.
        • Three officials of the municipal Fire Department visit the site.
        R: More than 10 housewives in the hamlet visit the mayor to complain about the danger and suffering.
        T: The director of City Planning Division and the chief of Landscape Unit of the municipal government conduct on-site consultation with local residents and Seibu.
        R: Some 20 resident representatives and the chairman of the Association for
The Protection of Greens and Historical Sites in Kamakura file a protest with Mayor Yamamoto.
        R: The Society in Defence of Hamlet S is officially launched.
Mayor Yamamoto pays an on-site visit.
        T: Tripartite talks between the municipal government, Seibu and local residents.
Mayor Yamamoto requests Seibu to conduct a strength test on the barrier and reconsider the green belt plan. Prefectural and municipal officials find illegal construction in an on-site inspection, and issue a cease and desist order (no notification to residents).

Table 8.2 (cont.)

1970	• The cliff at the back of the hamlet collapses due to heavy rain. Debris and fallen trees pile up in gardens.

1970 • The cliff at the back of the hamlet collapses due to heavy rain. Debris and fallen trees pile up in gardens.
R: Residents demand Mr. Tsutsumi, President of Seibu Construction, take safety measures against cliff failure.
B: Response from Seibu: "The cliff failure is unrelated to the construction works".
R: Unauthorized works are found from drawings related to the permission of Seibu's housing development projects.
R: Indictment of Seibu in the name of the chairperson of the Association in Defence of Hamlet S and the City Councillor.
R: A petition is filed with Mayor Masaki, calling for immediate cessation of unauthorized housing development works.

1971 Massive irregularities are found in Seibu's construction works. Kamakura Police Department sends papers to the prosecutor's office.
Without notifying local residents, the prefectural government permits the construction works to proceed, subject to conditions including the restoration of landscape on illegally developed zones.
R: A list is submitted to Seibu Construction describing damages suffered following the construction of the barrier.
B: An agreement on the indictment is reached between residents and Seibu: Seibu promises to hand over a park.
T: Opening ceremony for the park, which is handed over from Seibu to residents, and then to the municipal government.

2. Movement for the protection of Hiromachi Green

1970 Zoning regulation defines urbanization promotion areas and urbanization control areas. Hiromachi is designated as an urbanization promotion area reserved for first-class dwellings.

c1970 • Development projects emerge for Daimine and Hiromachi. Urban sprawl expands.

1973 D: A development project for Hiromachi is announced by a developer. The project is discontinued in the face of protests by local residents.

1978 The Basic Land Use Plan of the prefectural government designates Hiromachi as a "forest area".

1983 D: Three developers submit a development project application for preliminary review.
R: 60,000 signatures are collected against the development project.
The mayor puts the development procedure on hold.

1984 R: Local resident associations form the Joint Association for the Protection of Kamakura's Natural Environment. (Shichirigahama Neighbourhood Association will participate in 2000.)

1986 The Kamakura City Green Conservation Fund is launched.

1989 In his statement at a municipal board meeting, Mayor Nakanishi expresses his intention to allow developments in Hiromachi and Daimine.

Table 8.2 (cont.)

1990	D: The developers present a draft land use plan. R: The Joint Association hands over 120,000 signatures to Kanagawa Prefecture, calling for the designation of Hiromachi as a landscape area.
1991	D: Briefing starts for local residents (21 briefings up to March 1992).
1993	Newly elected Mayor Ken Takeuchi announces a revision to basic policy in favour of conservation.
1994	The municipal government refuses to accept the application for preliminary review submitted by the developers. D: The developers reject the municipal government's request for conservation, and try to submit an application for approval/consultation on the development project. The municipal government refuses to accept the application.
1995	R: 220,000 signatures are handed over to the City Council, calling for a municipal Landscape Conservation Ordinance.
1996	The municipal government rejects the application for development permission. D: The developers file a request for reconsideration with the prefectural Development Review Board.
1997	Kamakura City Council passes the Landscape Conservation Ordinance. The prefectural Development Review Board adjudicates the rejection of development application by the municipal government to be null and void.
1998	R: The Joint Association for the Protection of Kamakura's Nature launches a trust movement. R: The Association for the Protection of Nature in Kamakura-Hiromachi and Daimine is established (Director: Hisashi Inoue).
1999	The municipal government announces a Health Road Plan and the intention to maintain and conserve Hiromachi as a conservation-oriented city park. D: The developers reject a request from the municipal government including a natural environment study, and suggest demanding compensation.
2000	Following a recommendation of the municipal board, the municipal government adopts a basic policy for conserving Hiromachi as a "municipal forest".
2001	D: Two of the three developers, excluding Yamaichi Tochi, express their willingness to accept the offer of the municipal government to purchase 18 ha of land for Yen 11 billion. R: Three civic groups submit a request for cooperation to Mitsubishi Trust and Banking Corporation, the main financing bank for Yamaichi Tochi.
2002	• Mayor Tokukazu Ishiwata creates a Special Unit for Hiromachi and Daimine Greens. D: The developers renounce the project and enter into an agreement with the municipal government for full conservation.

Source: Created by the authors.
Note: "D" concerns developer; "R" concerns local residents; "B" concerns both developer and local residents; "T" concerns developer, residents and local government.

people must realize that they have a say in community matters and become actively involved in decision-making as a common practice. Only then can we hope for the establishment of a sustainable urban society, in terms of both the global environment and social viability.

Notes

1. Interview published in *Journal of Civil Service Exam*, June 1999.
2. A total of 337 environmental conservation groups responded to a survey conducted in the fiscal year 1997–1998 covering the 495 groups in Kanagawa Prefecture identified by the Environment Department of the Prefectural Government (March 1999). The 337 groups were distributed to individual municipalities (according to the mailing address of each group) and Table 8.1 shows the data for the top five cities in terms of the number of groups per 10,000 residents as well as for the two most populous cities, Yokohama and Kawasaki.
3. Other major objectives include "health care or welfare" (75 groups), "culture/arts/ sports" (76 groups) and "community building" (78 groups).
4. In Japan, the fiscal year runs from 1 April through to 31 March of the following year.

REFERENCES

Environmental Protection Association "Appeal from Kamakura" (1997) *The Old City of "Kamakura" is Now Under Crisis*, No. 17 (text in Japanese). (http://plaza.harmonix.ne.jp/~kamakura/japan/japan17.html)

Fujii, Keizaburo (2000) "New innovative Kamakura for all ages", Governing Board of Kamakura NPO Support Center (ed.) *White Paper on Civic Activities in Kamakura: Winds and Tides – Community and Civil Society for the Future*, Gyosei (text in Japanese).

Fujimura, Miho (2001) "What constitutes everybody's property? Land and People in a Village", Makoto Inoue and Taisuke Miyauchi (eds) *Sociology of Commons: A Reflection on Cooperative Management of Resources in Forests, Rivers and Seas*, Shinyosha (text in Japanese).

Funabashi, Harutoshi (2001) "Sociological research on environmental problems", Nobuko Iijima, Hiroyuki Torigoe, Koichi Hasegawa and Harutoshi Funabashi (eds) *Environmental Sociology in Japan. Vol. 1: Perspectives of Environmental Sociology*, Yuhikaku (text in Japanese).

Hasegawa, Koichi (2001) "Environmental movement and environmental policy study", Koichi Hasegawa (ed.) *Environmental Sociology in Japan. Vol. 4: Dynamism of Environmental Movement and Policy*, Yuhikaku (text in Japanese).

Iijima, Nobuko (1995) *Invitation to Environmental Sociology*, Maruzen.

Kamakura Citizens Round Table Conference on Community Building (ed.) (1999) "Community Building by Citizens: 27-year history of Kamakura Citizens Round Table Conference on Community Building", Ochanomizu Shobo (text in Japanese).

Kamakura City (1993) *The Third Comprehensive Plan of Kamakura City: Basic Strategy*, Kamakura City (text in Japanese).

—— (1994) *History of Kamakura City: Modern History*, Yoshikawa Kobunkan (text in Japanese).

Kihara, Keikichi (1998) *National Trust* [New Edition], Sanseido (text in Japanese).

Komai, Hiroshi (1985) "Creation of urban space and organization of residents: A comparison of Musashino, Kamakura and Narashino", *Journal of Sociology* 10: 99–108 (text in Japanese).

Matsushita, Kazuo (2002) *Environmental Governance: Roles of Citizens, Businesses, Municipalities and Government*, Iwanami Shoten (text in Japanese).

Miyamoto, Kenichi (2000) *Possibility of Japanese Society: Toward a Sustainable Society*, Iwanami Shoten (text in Japanese).

Nakazawa, Hideo, Naoyuki Mikami, Ken Ohori and Atsuo Terada (2001) "Is local environmental governance possible? The case of Kamakura City", *Social Information* 11(1): 113–26 (text in Japanese).

Nitagai, Kamon (1976) "Theoretical challenges and prospects for resident movements", Jiro Matsubara and Kamon Nitagai (eds) *Logic of Resident Movements*, Gakuyo Shobo (text in Japanese).

Sawa, Juro (1965) *50-year History of Kamakura Dojin-kai*, Kamakura Dojin-kai.

Seki, Reiko (2001) "Concept of environmental right and movements: From 'environmental right for resistance' to 'environmental right for participation and self-governance'", Koichi Hasegawa (ed.) *Environmental Sociology in Japan. Vol. 4: Dynamism of Environmental Movement and Policy*, Yuhikaku (text in Japanese).

Study Group on Environmental Rights, Osaka Bar Association (1971) "Recommendations for the establishment of environmental rights", *Jurist* 479: 60–84 (text in Japanese).

Terada, Ryoichi (1998) "Environmental movements and environmental policy: A Japan–US comparison on institutionalization of environmental movements and grass-roots democracy", Harutoshi Funabashi and Nobuko Iijima, (eds) *Sociology in Japan 12: Environmental Sociology*, Tokyo University Press (text in Japanese).

Torigoe, Hiroyuki (1997) *Theory and Practice of Environmental Sociology: From the Standpoint of Life-Environmentalism*, Yuhikaku (text in Japanese).

Yamamoto, Setsuko (1997) *Kamakura: A Seibu Kingdom*, Sanitsu Shobo (text in Japanese).

9

Housing complex replacement and "mental" sustainability

Yoko Shimizu

1. Characteristics of an urban housing complex and its residents

Cities have different types of condominiums, and in some areas, high-rise dwellings are becoming even taller, as if the only way is "up". The accompanying rapid growth in population has had various effects on local communities.

High-rise condominiums have evolved to meet the diversified requirements of residents, including greater rationality and functionality, increased security and protection of privacy. In particular, city-oriented people tend to prefer newer and higher dwellings.

Through replacement, housing complexes are improved and upgraded, but does this also improve the mental well-being of inhabitants, in terms of good relations among neighbours and a strong community network? Is it possible to sustain such emotional ties even after replacement?

When a run-down housing complex is replaced, the residents must choose either to move away or to move back into the new dwelling after it has been built on the same site. The replacement may also attract new residents from other areas. To those who live in the community, the replacement therefore is not only a change of house in which they live, but also brings various changes in mental ties in the community.

In order to examine the impact of urban condominium replacement on community life and the sustainability of traditional neighbourhood relations and community networks after replacement, this chapter presents a case study of Housing Complex A, a block of low-rise public apartment houses in the Tokyo metropolitan area that is about to be replaced. Through this case study, this chapter seeks to identify proactive measures needed to sustain or develop the stable livelihoods and community networks required for friendly relations and mutual support.

1.1 Overview of study area and characteristics of residents: ageing of houses and population

Housing Complex A is one of the largest public housing blocks that were built more than 30 years ago. With a population of 11,000 and about 5,000 households, the housing complex accounts for some 18 per cent of the municipal population. It is located in a leafy suburb, about one hour's commuting distance to the city centre. Since the opening of a monorail line several years ago, the area has become increasingly urbanized with a wave of condominium construction. The inhabitants of Housing Complex A, which was built as a public housing project, must meet certain income conditions, so the community includes more households on welfare or exempt from taxation than in neighbouring districts. There are also relatively many single-parent families, older single-person households and people with mental disorders, intellectual disability or suffering from incurable diseases.

The housing complex is currently undergoing a major rehabilitation. The more recent five-story apartment houses will not be replaced, but a living room and bathroom will be added to each apartment. Elevators will not be installed due to structural difficulties. Most of the housing complex, excluding the five-story apartment houses, will be replaced with high-rise condominiums with bathrooms and elevators.

The housing complex has aged with time, and so have the initial residents. Since the apartments were not designed for multigenerational cohabitation, children have had to move out after growing up in order to raise their own families. As a result, the number of older-person households is rising, including single-person households. Indeed, the district has one of the highest rates of ageing in the city (see Figure 9.1). This phenomenon is commonplace in urban public housing complexes. Could this problem have been foreseen over 30 years ago when the complex was built? In Housing Complex A at present, the apartment houses may have satisfied the needs of the initial inhabitants, but will not meet the needs of future generations. In this sense, the housing policy was not sustainable.

As houses age with time, the residents inevitably become older and the families change their lifestyles. Therefore, any housing policy should consider a time span of several decades, particularly when planning a hous-

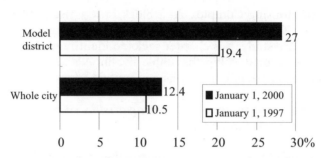

Figure 9.1 Share of the elderly in model district (Housing Complex A) (*Source*: Community Health Activity Report, Tokyo Metropolitan Government, 2000)

ing complex in an urban area. However, the government and local community usually do not begin to consider appropriate measures until actual problems emerge.

Problems facing a housing complex are not limited to personal matters, but often need to be addressed as challenges for the community as a whole (Housing Complex A in this case). Therefore, residents need to work in collaboration with the local government and other competent authorities to solve such problems.

More than 30 years after construction, Housing Complex A also faces various problems such as ageing. In order to address those problems, residents are working with the local government and related organizations to promote community building.

The residents of the housing complex are now waiting for their houses to be replaced with a mixture of hope and anxiety. As the replacement plan makes headway, anxiety seems to overwhelm hope. Concerns are mounting about the impact of the expected changes in residential environment on their actual livelihoods. In particular, local government officials share the concerns of the residents about the possible adverse effects of the new high-rise condominiums on traditional relations between neighbours and the community network that has been developed over the years.

Against this backdrop, residents have been working with local authorities (public health centre and municipal government) to identify and solve existing and potential problems in Housing Complex A.

1.2 Administrative response

With the promulgation of the Community Health Law in April 1997, a public health centre was established for the housing complex in an effort to reorganize the municipal public health system. This public health centre designated Housing Complex A as a model district to consider measures to support livelihood and promote health in collaboration with the municipal government, organizations and local residents. The former health consultation centre in charge of the district already knew that many of the residents had difficulties in living conditions or health.

The first step taken by the public health centre was to prepare and launch a three-year programme for community healthcare (April 1997 to March 2000), in order to identify problems relating to the residents' living conditions and health issues, their vision for the future of the district and support measures required to materialize that vision.

The objectives of the community healthcare activity programme (referred to as the "programme") were:
(i) to identify the needs of local residents and the conditions specific to apartment houses in the model district (Housing Complex A) and to establish the policy and measures to be followed by the public health centre; and
(ii) to develop a comprehensive support structure based on cooperation between local residents and healthcare, medical care and welfare organizations.

The primary focus of the programme was (i) coordinated efforts by the public health centre as a whole, (ii) collaboration with local residents and organizations, and (iii) direct feedback from the residents.

The programme comprised four elements:
1. Questionnaire survey in relevant organizations;
2. Mental health case study;
3. Everyday consciousness survey; and
4. Support for setting up a community gathering on health promotion and livelihood improvement.

Each component is outlined as follows:

(1) *A questionnaire survey in relevant organizations* is a method of identifying the needs of the residents through the relevant organizations. The objectives of the survey were: (i) to identify the general conditions of the model district and issues relating to the residents' health and living-conditions, (ii) to identify problems for relevant organizations and local welfare commissioners, as well as their requests to the public health centre, and (iii) to consider a comprehensive support structure for the health and welfare of the residents. The survey was conducted through interviews based on a questionnaire in 22 relevant organizations and 18 local welfare commissioners in the model district.

The survey identified general problems such as medication and diet (including nutrition) of the mentally disabled, health management of the demented elderly, the problems of the elderly living alone, domestic violence and child abuse. In addition, specific problems were identified such as: an elderly parent with dementia, who was the carer of a mentally-disordered child; the hygiene problems of an elderly alcoholic, including accumulation of garbage; neighbourhood problems caused by a mentally-disabled person. Particularly demanding challenges for professionals working for the organizations included intervention in cases involving those less than 65 years old who are not covered by existing services including the long-term care insurance system, as well as night-time emergencies of those living alone and increasing cases of senile dementia. The findings indicate that most of those cases are extremely difficult to address and that many families are facing multiple problems.

(2) *A mental health case study* to identify relevant needs by analysing specific cases involving a public health nurse, including mental health counselling. The survey was designed (i) to identify characteristics and needs of mental health cases in the model district, (ii) to review and examine the collaborating conditions among relevant organizations by systems and terms of mental health activities, and (iii) to review the policy and organization of the public health centre's activities by examining specific cases and objectively evaluating how problems have been resolved with the centre's support.

The subjects of the survey were 34 mental health cases in the model district that involved public health nurses working for the centre from April to December 1997. Survey items included: (i) background, attributes, family, lifestyle and medical history and conditions at the start of counselling; as well as (ii) living support needs ("Support requirements for the community life of persons with mental disorders and evaluation criteria" as defined by Masami Miyamoto et al. (1993)), interveners, means and frequency of intervention, and issues in providing support. The evaluation was made by measuring changes before and after intervention.

The survey revealed that there was little family support even though 80 per cent of the persons lived with their families. Indeed, support for the whole family was also required in almost half of the cases. Many of the persons with mental disorders did not have any support with their treatment or daily lives, and there were few opportunities to get to know their neighbours. In particular, many experienced "difficulties in living", requiring living support as well as health care. Since there are not sufficient social resources in or around the model district to provide living support, the study recommended that facilities such as "a living support centre" be built.

(3) An *everyday consciousness survey* is designed to collect information through interviews with selected groups. The technique of group interview, a type of "in-depth interview", gathers qualitative information (Ta-

dao Takayama 1998). The survey aimed at identifying the perception of the residents about the present state of the community and invited them to suggest ways to improve living conditions in the district, which would be taken into account in promoting community healthcare services in the years ahead.

Subjects of the survey were three groups: (i) resident representatives (leaders of the resident associations in the model district); (ii) local welfare commissioners; and (iii) the staff of the public health centre. Based on the interview guidelines described below, a moderator collected information through discussions. Information on linguistic and non-linguistic communication obtained from audio-visual recording and participatory observation was analysed by various experts including researchers and the programme staff. The analysis is based on qualitative research techniques such as descriptive, content, substance and relational analyses.

1.3 Interview guidelines

(1) Current status of the community and future challenges
(2) Concerns about life and health
(3) Image of an ideal community
(4) Realities of, and support for, residents with livelihood and/or health problems
(5) Issues related to services that have been used so far or will be used in the future
(6) Community ties and support network
(7) Healthcare and welfare support system
The results of the survey are listed below.

1.4 Current status of the community and future challenges

According to the survey, the positive aspects of the community include its natural environment, friendly relations with neighbours and the cohesiveness of residents. The following answers were obtained:
• "The air is cleaner than in the city centre."
• "Close personal relations with neighbours is a typical benefit of traditional terraced houses."
• "Terraced houses create a friendly neighbourhood."
• "It is easy to identify with any resident of the housing complex, even if he or she is a stranger."
• "In the local shopping area, I can buy large quantities of vegetables with another resident, even if he or she is a stranger."
• "Residents are willing to help one another if something happens."

- "Tea parties and trips are organized for older persons."
- "I do volunteer work outside the resident association."

On the other hand, residents voiced concerns about the imminent replacement, which might undermine the friendly relations and close ties
that have been developed in the community over the years.

- "After the replacement, I will not be able to step into my neighbour's
  garden to see if they are all right. The prospect of changing living conditions scares me."
- "I feel uncomfortable and uncertain at the thought of moving into a
  new house."

Moreover, some pointed out inconveniences concerning transportation.

- "There's no railway station close by. Although a monorail line was
  opened, it takes several minutes to get to the station."
- "Buses stopped running when the monorail line opened. It costs me almost ¥4,000 one-way just to go to a healthcare facility for the elderly or
  to see the family doctor by train and by bus."
- "I can take a short cut inside the housing complex by bicycle, but it will
  be harder as I get older."

The results show that the availability of transport services may have a
substantial impact on the medical care and well-being of older and ailing
persons.

Finally, some experienced difficulties in inter-generational relations
and community activities due to ageing as well as to changing lifestyles
and values of the residents.

- "I used to chat with other residents in the park with my young child.
  I never thought I'd feel so lonely now the child has left."
- "Community programmes cannot be implemented as young mothers
  are working outside the district. Children do not participate in community events, either."
- "One apartment house in the housing complex does not have any children under 18. I am always worried about how the residents' association will be run as people get older."

1.5 Concerns about living and health

With regard to living and health conditions in the community, residents
particularly voiced concerns about the well-being of single older persons
and the health-deterioration problems.

- "An elderly person living alone who fell down stairs broke a bone,
  couldn't move and was discovered three days later. Actually, I knocked
  on the door, wondering why the light had been left on all night. There
  was no stair rail."
- "I found a single elderly person lying down on the way to the garbage

collection site, and called the police for help. Later, I consulted with a social service worker about the person who did not look well, and the doctor found a bone fracture."

- "Many single elderly people keep a pet, which is not allowed, causing management problems."
- "Some elderly peoples' households are causing problems to neighbours due to foul odours and cockroaches, where they keep pets or cannot keep their rooms clean."
- "Many elderly people live alone. I'm taking care of an ailing elderly person who lives alone."

To address the problems of elderly persons living alone, some people suggested preventive action and early response by neighbours and local residents (resident association leaders and local welfare commissioners).

- "I suspect something has happened if the light is left on."
- "For families subscribing to a newspaper, a stack of delivered papers in the doorway indicates a problem."
- "The monthly clean-up day is the only opportunity to see one another, but many people don't show up on these days. Whether they participate in the clean-up or not, we would like more residents to show up so that we can see they're doing OK."
- "I want to check each day that people are well. I can't bear to think about elderly people dying on their own."
- "Older persons account for more than half of the residents of the apartment house that I live in. I did a training course for grade-three helpers to help older persons in cases of emergency. I am thinking about registering myself as a helper because that will allow me to help people in many ways including with the shopping."
- "Local welfare commissioners know who to contact in the case of emergency, but ordinary people don't. When a problem occurs, a neighbour will get in touch with a local welfare commissioner."

Thus, diversified support is needed to address problems the community residents face. Such support is actually provided for those in need by various people including direct neighbours as well as resident association leaders and local welfare commissioners.

1.6 Image of an ideal community

The following views were expressed on the image of an ideal community.

- "A community where elderly people can live in peace and security should also be a liveable community for other residents."
- "Community building should consider the convenience of the elderly."
- "An ideal community would not allow older persons to die in solitude."

- "In an ideal community, the elderly would be able to keep in touch with other residents."
- "An ideal community would accommodate people suffering from dementia or other health problems, rather than segregating them from the community by institutionalization or hospitalization."

Some were actually working towards those goals, while others emphasized the importance of mutual help among the residents.

1.7 Evaluation of services and information and related issues

As regards existing services, residents wanted greater user-friendliness.
- "The existing emergency call system will not work unless older persons understand how to use the equipment."
- "It is impossible to read the instructions without wearing glasses."
- "Phones are useful for communication, but the telephone is often out of reach in cases of emergency."

Some also suggested that the method of providing information should be changed, and that information should be provided as necessary.
- "Use larger letters, illustrations and underlines in the information leaflet to facilitate reading, otherwise people won't be bothered to look at it."
- "As I had participated in the Community Gathering on Health Promotion and Livelihood Improvement, I thought of consulting the public health centre about my expectant daughter. The public health centre responded quickly, but I would never have thought about using the centre without any knowledge."

1.8 Measures and challenges for solving problems

Concerning community ties, residents gave the following answers.
- "I had many contacts with local people 30 years ago, when the housing complex was built. Now I miss those relations."
- "There was a well-developed community programme when the housing complex was built. Nowadays, each apartment house acts separately. I want to build ties with local shopkeepers."
- "We are using a public bathhouse because there is no bathroom in the housing complex. The bathhouse gives us a chance to go out and get to know each other."
- "An assembly room is needed in the housing complex."
- "We organize concerts in cooperation with local schools."
- "I am wondering how we can make friends with older persons. There are many difficulties, including just how to address them."

The residents are concerned about the dilution of traditional close relations among themselves, but still admit that the attractiveness of the community lies in the "emotional ties and personal contacts that stand the test of time". The findings indicate that the residents are strongly in favour of continuing those community contacts after the replacement of apartment houses.

With regard to proactive intervention and support for residents with existing or future well-being and health problems (namely, the elderly and those undergoing medical treatment), more systematic efforts are required by the community to ensure that problems are quickly spotted and support is given to those who need it.

- "Every condominium should have a warden, as in the case of the special housing complexes for the elderly run by the metropolitan government."
- "Activities in the model district (community organization programme) should be extended to the whole city."

Although residents are trying to solve community problems by themselves, they are keenly aware of the inherent difficulties.

- "They say we have to help each other as we get older, but that's easier said than done."
- "Far more people should share the pleasure of participating in community gatherings."

Thus, the local government and professionals are required to support initiatives by local residents, so that community activities may survive the replacement of the housing complex.

Regarding the system of support for healthcare and welfare, residents called for the establishment of a reliable medical institution in the neighbourhood, as well as a special healthcare or medical care facility that can respond to emergencies.

- "I would like to receive treatment in a reliable medical institution."
- "There should be a public general hospital and a healthcare facility for the elderly in the district."
- "I need a reliable medical institution that can receive patients in emergency situations and at night."

1.9 Anxiety over the replacement

- "After the replacement, I will not be able to step into my neighbour's garden to see if he or she is all right."
- "The closure of the public bathhouse means the loss of the only opportunity to get in touch with local people."

Thus, the residents are concerned over the replacement of present apartment houses with high-rise condominiums. At present, no apartments have bathrooms, so the local bathhouse provides a place of contact be-

Figure 9.2 Community gatherings on health promotion and livelihood improvement (*Source*: Photos by the author)

tween local residents. Also, the present terraced houses have helped build friendly relations between neighbours. The expected transformation of the residential environment into the housing complex has raised concerns about possible changes in the traditional friendly relations between residents and potentially difficult relations with new residents. Some even suggested that the changing environment might precipitate dementia and isolation among the elderly, and increase the number of cases of "withdrawal".

(4) As older persons successively died in solitude in the district, the union of local resident associations adopted the campaign policy of "No More Deaths in Solitude" to show their commitment to building a caring community. The public health centre was also aware of the need to work with local residents and other organizations to promote community-building activities. This cooperation culminated in the organization of a Community Gathering on Health Promotion and Lifestyle Improvement.

The objectives of the gathering were: (1) to create a forum to discuss health problems and everyday life in the community with local residents and other organizations; (2) to collect information about the living conditions of individual residents, personal relations and atmosphere in the community and requests from residents for consideration in future programmes of the public health centre and organizations; and (3) to collaborate with residents and related organizations from the planning stage to encourage initiatives by the community. The gathering was launched in January 1998 with the participation of government officials and local residents, and has been held regularly ever since.

The theme of the first session was selected by the residents from several options presented by the public health centre. Resident representatives (leaders of resident associations) organized the meeting in consultation with public health centre staff and in coordination with organizations. The leadership of resident associations in organization and information activities has been established over time, and the gathering is now held at the initiative of residents. Similar gatherings are also organized at the level of apartment blocks.

1.10 A session of the community gathering on health promotion and lifestyle improvement

A session is organized in a participatory manner, beginning with small group meetings so that everyone can have a say, and ending with a plenary meeting. Various types of learning through experience and health education are also introduced. Residents who are not members of a resident association may also participate.

The gathering has been held under various themes including the following examples. The first session of the gathering included learning through experience to understand how an older person perceives the environment of the housing complex. It provided an opportunity to think about how to improve the residential environment and consideration for older persons.

1.11 Some themes of the gathering

1st session: Building a Friendly Community for Older Persons by Learning through Experience
2nd session: The First Step to Enjoy Eating
3rd session: Let's Talk about Dementia
4th session: Peaceful Life after Retirement
5th session: Stay Healthy Forever: Keys to Good Health

The gatherings have produced the following results.

(1) Despite the participation of resident representatives in the planning as facilitators, the public health centre took the lead in organizing the first session. Subsequently, however, the resident representatives became increasingly confident as they took responsibility for providing information and hosting the sessions. Most of them wanted the programme

to continue as they gained from tackling the difficult work, and they now take the initiative in organizing the sessions.

(2) The gathering has stimulated the sense of participation in health promotion and community building among local residents. By experiencing the lives of the elderly and participating in group meetings, residents now share other people's ideas and standpoints. "I realized that we need to be ready to help each other." "I now feel like participating in the community programme." "I would like to start by getting in touch with my neighbours." Those feelings expressed by participants after a session, included in a questionnaire survey, clearly indicate that the gathering has made a real difference. Moreover, both resident representatives and participants said that the exchange of views allowed objective analysis of their own situations and strengthened their commitment to building a liveable community themselves based on the awareness of common problems. Thus, the gathering is now functioning as a forum for discussing specific measures to achieve that goal.

The gathering has had significant ripple effects. Other communities and groups have requested the cooperation of the public health centre in organizing similar meetings, and the meetings also resulted in new consultation services by organizations. The gathering is expected to produce further results in the future, such as the empowerment of local residents and the promotion of health and community networking.

2. Hopes and fears for housing complex replacement

The findings of the survey and activities described in the previous section may be summarized as follows.

2.1 Emotional ties in the community

Residents regard close personal relations, the small-town atmosphere of terraced houses and friendly neighbours as "merits of the community". Since it was built, Housing Complex A has never had bathrooms or elevators. The oldest apartments in the complex are wooden two-story houses as shown in the picture; their steep stairs are dangerous for older persons and many have tumbled down them and broken bones. Nevertheless, wooden apartment houses allow friendly ties between neighbours. Noise from a neighbour means he or she they are all right, and a shout to a neighbour will be heard. They can also chat together in the garden.

In particular, residents praise the terraced (two-story) style of the apartment houses for developing community cohesiveness through

Figure 9.3 An old two-story apartment house (*Source*: Photo by the author)

friendly relations between neighbours. Since no apartment has a bath-room, the public bathhouse provides a good opportunity for chatting among residents, as well as for checking that someone is all right, espe-cially for those who tend to stay indoors.

As the replacement plan made headway, however, residents increas-ingly expressed the concern that such "merits of the community" might be lost. The replacement of existing apartment houses with high-rise con-dominiums will lead to material improvements such as bathrooms, but it has been decided that the public bathhouse will be closed. Furthermore, residents are concerned about relations with newcomers. Although their new houses have not been decided yet, current residents are aware that relations with new neighbours might never be the same again.

2.2 Living and health problems

Multiple surveys have shown that many residents are facing living and/or health problems, which sometimes lead to emergencies. For example, one older person living alone fell down the stairs, could not move, and was

Figure 9.4 A five-story apartment house (extended) (*Source*: Photo by the author)

discovered three days later. Ailing residents often postpone treatment until they have to call an ambulance in an emergency. Many older persons keep a pet, which is against the rules. Inappropriate care for pets causes hygienic problems including unhygienic pests and nuisance to neighbours.

Many of the older persons had previously lived with their families, but were forced to live alone as the composition of their families changed over time. Although the issues of societal ageing should be addressed at the national level, the government has not developed sufficient measures to support or protect the elderly population who, as in the case of Housing Complex A, manage to live on their own but are vulnerable to sudden deteriorations in health and living standards if they fall or are taken ill.

For the withdrawn or slightly demented elderly people who manage to live on their own, the large change in residential environment to a high-rise condominium will have a major impact on their lifestyle, including the loss of friendly neighbours, separation from pets, rebuilding of rela-

Figure 9.5 High-rise condominiums (being newly built) (*Source*: Photo by the author)

tions with newcomers, the cumbersome task of using the elevator to put out garbage, and the difficulty in asking a neighbour for help. In other words, a change in residential environment and the configuration of the community requires the residents to adapt to the new environment in every aspect – physically, mentally and socially. In particular, the unfamiliar lifestyle of residing on higher floors and the closure of the bathhouse might limit the scope of activities for older persons and those undergoing medical treatment, thus exacerbating their isolation.

Continued efforts will also be required even after the replacement takes place in order to address the issues that have already emerged, such as the problems of elderly live-alones, and the frequent use of ambulances in emergency situations caused by belated treatment. The local government should therefore work with residents to develop appropriate measures for building an ideal community, including the establishment of a medical institution, the prevention and early detection of ill health and emergency response, so that the housing complex, after replacement, can continue to accommodate elderly live-alones and patients undergoing medical treatment.

2.3 Community support system

Many people in the community are facing various problems and require support, and there are residents who actively try to support those people. Such problems are diverse, complex and often serious. Many of them cannot be solved by the residents alone and hence require professional assistance.

Specifically, residents want a reliable medical institution nearby and a joint forum for liaising with the government, including local welfare commissioners, the municipality and the public health centre. In their view, "it is important that people have opportunities to share information and problems in difficult cases". As the following remarks show, resident association leaders and local welfare commissioners are striving to develop a network of support for those who need collaboration with other residents and health promoters in nearby communities:

- "I did a training course for third-class helpers so that I can help in an emergency."
- "A local welfare commissioner is helping a person who became blind due to diabetes and whose family includes a person with intellectual disability."
- "I took care of an elderly live-alone in consultation with a helper, starting out by helping with meals."
- "We organize bus trips for the elderly twice a year. People in wheelchairs also participate."
- "Promoters of the community welfare organization programme take care of older persons living alone. This model programme should be extended to the whole city."

On the side of a support party, the public health centre staff made the following remarks:

- "I used to help with meals for the elderly at the community hall. The activity ended with the death of the leader, who was also elderly."
- "People are actually helping each other. For example, neighbours and local welfare commissioners accompany those who apply for registration as patients with incurable diseases. The problem is that an increasing number of people require such help."
- "We should recognize that many residents are ready to help others. A mechanism is needed to allow people to work for others as well as for themselves."
- "[It is important] to empower the community by finding what local residents can do without relying on services provided by public and private entities."

Those findings indicate that resident association leaders and local welfare commissioners have an important role to play as a support party in cases that require community support. In addition to improving direct personal services, the government needs to consider arrangements for professionals to support such carers within the community.

3. Mental adjustment and barrier-free environment

3.1 Adaptation to changing residential environment and required support

The housing complex replacement will transform most of the existing apartment houses into high-rise condominiums. Inhabitants will therefore be required to rebuild the community, including their relations with neighbours and other local residents that have been developed over the years. Although the residential environment will improve, they are deeply concerned about possible changes in life patterns, such as relations with newcomers, further withdrawal of older persons and the loss of human contact with the closure of the public bathhouse. Concrete measures to address those issues should therefore be considered, as well as a system to provide support when problems emerge. Preventive measures would include setting up a community centre to take over the role currently played by the bathhouse, and to enhance the caring activities undertaken by neighbours and local welfare commissioners. Any such measures will require collaboration between residents and the local government.

   In order to ensure early detection of, and response to, problems in any residential environment, the convenience and effectiveness of existing services need to be reconsidered from the viewpoint of ordinary people, that is, the users of services and their neighbouring carers. Major challenges for professionals include providing direct support for those in need, facilitating caring activities by residents themselves and building a mechanism to solve problems in collaboration with the community (see Figure 9.6).

4. Community networking and "mental" sustainability

4.1 Image of a community network

Various studies on the district and activities conducted in collaboration with local residents have shown that support parties in the neighbour-

---

Residents' image of an ideal community

"Friendly and livable community for the elderly and the disabled"

Issues

Issues for the whole community
1. Increasing problems related to the health and living of (single) older persons as ageing progresses in the model district.
2. Concerns about the collapse of local communication following the replacement of apartment houses.

Issues for support parties
1. It is difficult to intervene in cases involving persons not covered by social services, often resulting in emergency response.
2. It takes time to share experience in difficult cases.

Viewpoints and measures for solving problems

Residents
◎ Sustained communication and effective utilization of various social resources
1. Stay fit by taking care of one's health in order to lead an independent life.
2. Maintain communication with neighbours and friends.
3. Find out about social resources that can be utilized in case of emergency.

Healthcare centre staff
◎ Local coordination by appropriately identifying needs
1. Recognize the centre's role in the community.
2. Maintain close contact with local key coordinators to ensure effective assignment of roles.
3. Provide summarized information on community healthcare and welfare.
4. Promote health education.

Other organizations
◎ Building a system that ensures smooth transition from counselling to support
1. Identify the needs of residents and share information among them.
2. Promote the community welfare organization programme implemented by the social welfare council.
3. Create appropriate forums including liaison meetings.

Focus of collaboration between
the government, private entities and residents

"Mutual understanding and information sharing" is the key to enhancing collaboration.

Figure 9.6 Viewpoints and measures for solving community issues (*Source*: created by the author)

252

hood, such as resident association leaders and local welfare commissioners, are taking the initiative in building a friendly community for the elderly and the disabled. With the ageing of the community, however, support is needed by an increasing number of residents, including elderly live-alones as well as families facing multiple problems such as dementia, isolation and incurable diseases.

Many of the problems cannot be solved by professionals or neighbours alone. Any deterioration of the situation, such as immobility, will make it difficult for the person to live in the community unless he or she can receive support from a neighbour. In order to prevent deaths in solitude, residents want a community network by which any signs of difficulty or an emergency for live-alones can be immediately detected. It is essential that local authorities and other organizations exchange information and cooperate with community representatives, such as resident association leaders and local welfare commissioners, as well as with other residents to build such a support network. Close relations between neighbours and strong community ties will provide the basis for the network.

In this context, Figure 9.7 shows the concept of an ideal community network for Housing Complex A.

The following measures will be needed to ensure that all parties comprising the network assume their respective responsibilities in networking activities to ensure the network works effectively (Figure 9.8).

4.2 Measures to build a network and enhance its functions

(1) Providing residents with easily understood information on health and lifestyle (reviewing health education and public relations magazines);
(2) Ensuring that residents know who to contact and clarifying the roles of each agency;
(3) Mutual feedback of information between residents, local authorities and other organizations;
(4) Review of the proactive support network and the emergency support system;
(5) Training of and support for resident carers and community leaders for health promotion;
(6) The involvement of residents and professionals across a wide range of fields in order to reflect the needs of all age groups living within a community-style building;
(7) Setting up a structure to ensure collaboration between residents and professionals in building an ideal community (e.g. Community Gathering on Health Promotion and Lifestyle Improvement)

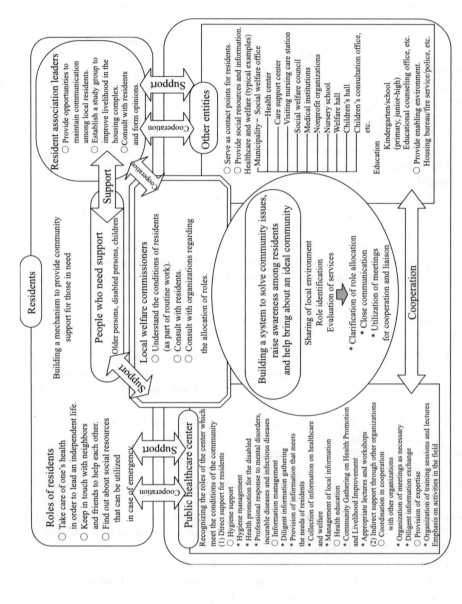

Figure 9.7 Image of a community network for Housing Complex A (*Source:* created by the author)

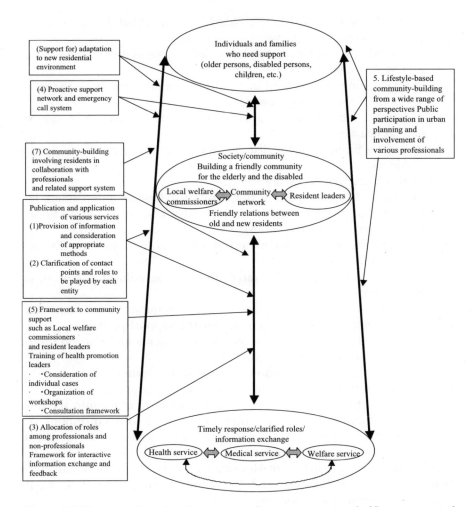

Figure 9.8 Future tasks related to community support network (*Source*: created by the author)

5. Future prospects

The following challenges will have to be met in order to create an ideal community in Housing Complex A.

(1) In respect of health and everyday problems for the elderly and disabled, it is necessary:

(i)   to improve the living environment and mutual help between community members so that the elderly and disabled can go out and about in safety, thus preventing isolation;

(ii) to provide the necessary opportunities for maintaining communication among community members;

(iii) to take ageing into account in establishing a mechanism for the management of the housing complex;

(iv) to build a framework of support for elderly live-alones in case of emergencies;

(v) to build a framework for disease and accident prevention in ageing residents.

(2) For residents, it is necessary:

(i)   to take care of one's own health in order to lead an independent life (information to be provided on necessary knowledge, attitude and practice for health promotion);

(ii) to keep in touch with neighbours and friends to build mutually supportive relationships;

(iii) to find out about social resources that can be utilized in case of an emergency.

(3) In creating a support system for solving community issues and building an ideal community, it is important for the support parties to:

(i)   build a system to identify the needs of residents and share relevant information among those involved;

(ii) build a system of cooperation for supporting residents;

(iii) evaluate services in order that they may be improved.

The residents of Housing Complex A have begun to move into the new high-rise condominiums, and the local authorities and organizations are working with the them to address the concerns that have emerged with the housing complex replacement.

It is hoped that the merits of Housing Complex A – namely, the emotional ties, friendly relations and community network – will be maintained and enhanced following the replacement.

REFERENCES

Miyamoto, Masami, Tanaka, Naoki and Kinoshita, Masanobu (1992) "Support Requirements for the Community Life of Persons with Mental Disorders and Evaluation Criteria", Report on the 13th Annual Meeting of the Japan Society of Social Psychiatry (text in Japanese).

Murayama-Yamato Public Health Centre, Tokyo Metropolitan Government (2000) *Toward a Comprehensive Framework of Support for Community Health Needs: Start from Seeing. Community Health Activity Report* (text in Japanese).

Shimizu, Yoko (1998) "A study on the evaluation of public health nurses in community health activities: With focus on livelihood support for users and cooperation", *Journal of Japan Academy of Nursing Science* 18(3): 428–429 (text in Japanese).

——— (2001) "A qualitative study on the needs and problems of condominium residents with group interview method", *Japanese Journal of Human Science of Health-Social Services*, 7(2), 43–52 (text in Japanese).

Shimizu, Yoko and Anme, Tokie (1999) "A study on community health activities and their evaluation: Group interview survey for local welfare commissioners", *Journal of Japan Academy of Nursing Science* 19 (text in Japanese).

——— (2000) "A study on the needs and evaluation of community health activities: From a focus group interview survey for public health centre staff", *Journal of Japan Academy of Nursing Science* 20 (text in Japanese).

Shimizu, Yoko and Moriizumi, Junko (1999) "Participative community health activities and their evaluation: From a vicarious experience session on older persons in the community gathering on health promotion and livelihood improvement", *Japan Journal of Public Health* 45(10) (text in Japanese).

Shimizu, Yoko and Suzuki, Akiko (2000) "A study on community health activities in condominiums and their evaluation: From a group interview survey for resident association leaders", *Collected Papers of the Japan Society of Nursing (Community Care)* 30: 92–94 (text in Japanese).

Suzuki, Akiko and Shimizu, Yoko (1999a) "Needs identification from community health activity programmes for building a comprehensive framework of support", *Journal of Public Health Nurses* 55(10): 829–835 (text in Japanese).

——— (1999b) "Participative community health activities and their evaluation: From the FY1998 session of the Community Gathering on Health Promotion and Livelihood Improvement (2nd Report)", *Japan Journal of Public Health* 46(10) (text in Japanese).

Takayama, Tadao and Anme, Tokie (1998) *Theory and Practice of Group Interview Method: How to Gain Insight from a Qualitative Survey*, Kawashima Shoten (text in Japanese).

10

Sustainable community development: A strategy for a smart community

Osamu Soda

1. Sustainable urban planning

Sustainable community development, as described in this chapter, is an offshoot of the concept of sustainable development. In particular, it may be considered as a product of community-level effort in Japan. According to the Bruntland Report (1987),[1] sustainable development "meets the needs of the present without compromising the ability of future generations to meet their own needs". The International Council for Local Environmental Initiatives (ICLEI 1994),[2] established by the OECD, defines sustainable development as the kind of development that supplies all members of the community with basic environmental, social and economic services without devitalizing the underlying natural, artificial and social systems.

The word "sustainable" has two connotations. One of them concerns the viewpoint of "metabolism", which requires that our generation does not tip the balance between resource production and consumption. Although it is improbable that the former exceeds the latter, the requirement calls for the minimization of the deficit. Particularly important is the idea of stopping the consumption of social resources entirely. According to this viewpoint, it is not sufficient to accumulate flows as stocks. Flows should only be distributed and consumed to the extent necessary to cover any deficit after making the most of the stock.

Another meaning concerns the viewpoint of "adaptability". Social phenomena form a "fabric". A series of adverse effects in an uncertain situation might tear the fabric by exerting an unexpected force. It is therefore necessary, if we are to create stable conditions that ensure the continuation of the present situation, to build a framework for averting any eventual risks. Thus, one of the major challenges for urban planning is to transform the existing community into one that satisfies the two requirements of sustainability. For the purposes of this chapter, the term "community" refers to any sphere that is wider than a group and smaller than a city. Although individuals may be identified to a certain extent at this level, the behaviour and network of individuals have to be distinguished from the mechanism of the whole society. In terms of space, a community ranges from several city blocks to a primary school district or a geographic division of a municipality.

The members of a "community" have a variety of views. For example, they differ in age, social class, social status, academic level, values, nationality, lifestyle, economic view and environmental perception. Being a collectivity of individuals, a community typically reflects the lifestyles and personalities of the residents. As part of a larger urban society, however, a community is also required to function in conjunction with and coordinate well with the overall mechanism.

The requirements of "metabolism" and "adaptability" entail a fundamental change in a community. The objective of sustainable community development will not be attained if existing organs simply continue to use the same forms and methods in conducting activities for environmental purposes, such as: recycling, clean water, tree-planting and biodiversity. Environmental sustainability will emanate from a completely renewed, sustainability-oriented social system. Such renewal will hinge on the following: self-determination by the community; transferring financial resources; self-management; transparent procedures; acceleration of decision-making processes; improved quality, user-friendliness and efficiency of plans and services; enhanced accessibility; control over overexploitation of resources; enhancement of oversight by competent authorities or third-party organs; strategic advice, and clarification of standards. This vision of a sustainable society is sometimes known as a "smart" or "enhanced" community.

In light of both the social system needed to achieve a smart community and current social conditions, this chapter seeks to provide the basic ideas necessary in defining the "system" and "design" of a social system, as well as their transition.

2. Incorporation of the ecological viewpoint into community development: a brief history

2.1 Some classic theories

The previous section referred to the viewpoint of "metabolism". A sociological approach brings a significant insight to this connection. By regarding the human community as an organic entity, this approach, known as ecological social theory, examines the interaction between the actor and the environment so as to consider how the "design", "system" and "behaviour" of the actor may change as a result. Human beings create a secondary social environment within nature and form a unique artificial ecosystem, which has a substantially different mechanism than that of other creatures. It may be said that the analysis of this human ecosystem originated in the *Migrationstheorie* put forward by biologist M. Wagner (1868). This idea was further developed by geographer F. Ratzel into a historical theory of Cultural Diffusion (1894) among peoples of the world. According to this theory, human beings are not simply regulated by the environment but respond to its influence and create a new cultural environment. He referred to a collectivity of the same cultural spheres as *Lebensraum*. Subsequently, sociologists R. Park and E. Burgess applied *Migrationstheorie* to explain urban structures through the development of a Concentric Zone Theory (1924).

It was P. Geddes who extended the scope of the ecosystem theory from the level of explanation to that of planning in order to discuss the behaviour of an actor responding to the environment. After studying biology under T. Huxley, Geddes made a career switch to urban planning study. He was involved in various research activities on both theory and practice in many parts of the world, including the UK, India and France. *Cities in Evolution*, published in 1913, was based on that experience. Adapting the Darwinian theory of survival of the fittest to urban studies, he argued that the force of urban (borough) reorganization, rather than militarism and industrial upheavals, would be needed in order to survive the age of competition among world cities, which dawned with the industrial revolution. Geddes advocated the study of "civics", maintaining among others that:
- it is necessary to observe and analyse citizens and cities as Aristotle observed organisms;
- new technologies should be used effectively to ensure order in cities;
- the development of cities and the formation of slums may be explained by a law;

- control over cities may enable the prediction and early solution of urban problems;
- it is urgent for that purpose to develop scientific research techniques and a theory of urban design;
- easy opportunities need to be created to educate citizens on urban planning, such as exhibitions, and
- the development of citizens themselves into able practitioners requires the establishment of research centres in cities, so that citizens may work with professionals when considering urban planning.

Supported by *Migrationstheorie*, C. Doxiadis subsequently gave a reasonable explanation of urban forms and land-use structures. In *Transitional Cities* (1963), he presented a theoretical model of urban growth and the transition of centres, noting that, based on a morphological logic, a specific environment is formed in every part of a city. Recognizing that cities may expand infinitely through rapid economic growth, Doxiadis saw the urban environment as a human settlement issue and played a substantial role in developing national policies and social systems by pursuing international dialogue. In this context, he advocated ekistics, a discipline that considers how cultural environment changes at each of the 15 levels of human habitation, including rooms, houses, colonies, small cities, large cities, and world cities.

In his work, *Operating Manual for Spaceship Earth* (1963), engineer-philosopher R. Buckminster-Fuller warned that misjudgement and unlimited desire might endanger the operation of Spaceship Earth, which has limited but sufficient resources for all mankind to live happily.

Biologist G. Hardin described how citizens could understand and overcome this problem. In his work *Filters Against Folly* (1985), he argued for the necessity for able citizens, not only experts, to make decisions, as well as the need to acquire at least three ways of thinking called filters, lest the judgement of citizens should be overwhelmed by complex realities. Those filters refer to literacy, numeracy and ecolacy.

The 1990s saw an increased search for a clue to establishing sustainability measures. Malthusian scepticism was replaced by environmentalism, focusing on human efforts to make real mankind's hopes for the future. The concept of "sustainable development" has rapidly become popular since the 1992 Global Environment Summit when US Vice President Al Gore advocated its mainstreaming at the level of international policy.

Based on the background presented in this section, the following sections discuss three system of "governance", "resources mobilization", and "feedback" as important "systems" which enhance the adaptability to new sustainable urban development.

2.2 Sustainability in organisms

In the first place, the formula of sustainability mentioned in Chapter 2 should be revisited in the context of community development.

What is the relationship between "metabolism" and "adaptability", essential elements of sustainability, on the one hand, and the "design" and "system" of a sustainable community, on the other?

For example, Geddes compared a human settlement colony to a coral reef, and described the emerging metropolitan areas as "conurbations". Tatsuo Motokawa (1992) explained why corals have come to adopt that exact "design" and "system", that is, the mode of life as an immobile animal.[3] A coral lives in symbiosis with a unicellular plant of zooxanthella, which produces the necessary nutrition by photosynthesis (self-sufficiency in energy). Its simple exoskeleton, in the form of brickwork, represents the optimal method of construction that allows a simple immobile thing to increase in size (replicability/stackability). A coral reef is a large colony of individual corals. Each individual replicates itself by cutting its body in half or shooting out sprouts from parts of its body. A replicated individual remains stuck to the parent at a soft part that also connects nerves. Unlike a vertebrate individual, swift, controlled actions are not possible. Nonetheless, a certain extent of coordination may be observed, as an attack on a polyp will cause neighbouring polyps to withdraw to the stone house (autonomous decentralized cooperativeness). Coral colonies are immortal (permanent cycle). Although old individuals die, the lime exoskeleton made of their secretions remains as "resources" for future generations, and grows slowly but infinitely, covered by new individuals (social capital formation). Thus, corals become a sustainable creature as a result of various characteristics of their design and system, such as self-sufficiency in energy at the level of the individual, simple replicability and stackability, flexible autonomous decentralized cooperativeness, permanent cycle, and social capital formation. Those features have some implications in terms of the human social system.

Others also explained the sustainability of a community or society when compared with an individual organism. For example, political scientist R. Putnam compared American capitalism to a saguaro (a kind of cactus). Thus, he launched a Saguaro Seminar, an intensive training curriculum for a next-generation participative society at the John F. Kennedy School of Harvard University. Despite often being detested or despised for its spiky appearance, the cactus actually serves as a measure of health of an ecosystem. It is a tough, long-living plant that grows little by little and can even survive in arid areas. It provides trunks for grapevines, nests for birds, and pleasant shade for many creatures. Native Americans are said to bless this plant.

2.3 Sustainability of the human society

It may be true that coral and cactus are highly sustainable organisms. However, no higher animals or human beings can follow their examples of remaining immobile, or transforming their appearances into simple and efficient shapes. Yet, as studied by biologist K. Schmidt-Nielsen, some desert animals, such as the kangaroo rat, have developed a sustainable water-saving bio-mechanism.[4] Bradypods, a poikilotherm, also minimize resource consumption.

In contrast, humanity and human society have evolved far more actively in their response to the environment. Joints that allow swift movements, the bone structure to support them, muscles to move them, homoiothermy to carry out functions at any moment, and the development of a central nervous system and cerebrum to issue instructions to every part of the body – all attest to the process of such evolution. Despite a considerable loss of energy, human beings have survived sudden and diverse changes in the environment by acquiring the necessary learning ability. We have explored problems under unknown and highly uncertain circumstances, pre-empted oncoming changes, and developed a capacity to adopt proactive behaviour. Individuals have grouped themselves to form a sophisticated and complex high-mobility society. Of the two meanings of "sustainability" described at the beginning of this chapter, it is "adaptability" that human beings have developed in order to survive environmental changes. Moreover, they have established a stable relationship with the environment, not by unilaterally adapting themselves to the environment but by changing the latter, as necessary, through interaction.

For example, Mitsuo Ichikawa (1982) clarified the original relationships of human beings with their houses and the environment through a study of the Mbuti Pygmies, an ambulatory hunting people in the African forests.[5] The pygmies build a camp on their way to hunting, each family making a hut with locally available materials. Once the camp is completed, they start hunting in the surrounding area. Subsequently, the hunting area is widened progressively as the density of game declines. Finally, when they realize that it takes too much time to go hunting, they move to another place. By refraining from taking all the animals, the pygmies incorporate consideration for environmental resilience into their hunting system. They use the huts – simple, temporary shelters – for sleeping only and abandon them as they move along. Their way of life is supported by a simple social structure and individuals who do not have more than they need.

To humanity and human society, sustainability means, in the first place, exploring at the social level how to ensure the survival of human beings

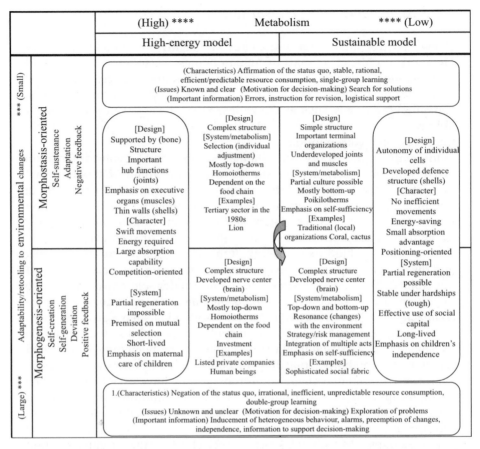

Figure 10.1 Sustainable "design" and "system" in terms of metabolism and adaptability (*Source*: created by the author)

themselves by working with the environment without destroying it, utilizing it and transforming it into their own environment. It can never mean to remodel us into some sort of energy-saving creature in response to the harsh environment, as with the kangaroo rat.

It goes without saying that our fundamental habits and ways of life require compelling changes, including mass production, mass waste and high energy consumption. There is no future for humanity without these changes. At the same time, however, the argument that "we need to start by changing our lifestyles to more ecologically-aware ones" will not necessarily be the driving force in achieving a sustainable society, even though it is correct in normative terms. If we consider human society as analogous to individuals, the mortal sin in individuals of the desire to use and then throw away will necessarily lead to a high-metabolism system of society. Thus, it is important to make intentional changes in this fundamental structure in order to bring about a society with a sustainable design and system.

What needs to be underlined here is the hypothesis that it is possible to advance the fabric of society, which represents a more complex organic system than the human body.[6]

Coral and cactus maintain a large extent of sustainability in them. As individual units and cells they embody a simple and highly efficient low-metabolism system. In contrast, mankind is required to retool the whole system into a sustainable mechanism by utilizing the sophisticated wisdom and energy of its individual members (cells). This relationship is shown on the horizontal axis of Figure 10.1 (see Section 5 for the description of the vertical axis).

3. Governance system

3.1 Changing institutional environments for communities

The design of an actor changes through "metabolism" and "adaptation". In this process, the actor is exposed to the force of the environment. The actor transforms itself in response to the surrounding conditions, that is, environment. Spinoza thought that, in terms of the relationship between the actor and the environment, the actor creates an environment in order to preserve itself, and not that the environment unilaterally controls the actor. What he called "conatus" is the will of the actor to improve its vitality, in part through adaptation to the environment. He argued that genuine happiness might be obtained by understanding natural order and controlling emotions. A shift to sustainability will call into question the whole social system as well as the behaviour of the actor.

In this context, the changing systemic (institutional) environment should be examined in relation to the design of society that conditions our efforts to build a sustainable community. The relevant concept is shown in Figure 10.2.

Various arguments exist on the influence of institutional forces on a community in making efforts to transform itself to a sustainable community.[7] If we define institution as the interaction between different sized clusters, including central government, regions, prefectures, municipalities and communities, as well as the prevailing social environment, the crux of the matter lies in how to systemize its design (structure) and system (rules).

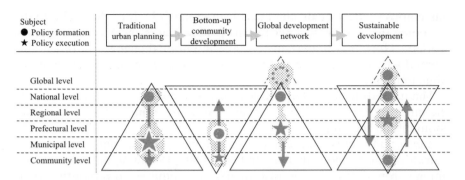

Figure 10.2 Changing institutional environment for policy-makers (*Source*: created by the author)

In traditional urban and environmental planning, the framework defined by central government is transmitted downstream to the prefectural governments, which are required to put in place the framework. Although some central government authorities have been delegated to local governments in recent years, the communities are still perceived as mere receivers of orders.

Since the 1980s, in particular, the bottom-up style of community development has been practised in Japan as an antithesis to that tradition. Municipal and prefectural governments have encouraged the autonomous execution of various initiatives at the community level. This approach is based on the so-called principle of subsidiarity.

In the late 1990s, when globalization was in full swing, sustainability became a global issue that prefectures or even countries could not address alone. It is now considered that any policy for sustainability should cover the entire environmental unit that constitutes a biosphere, including: baysides, river basins, climatic zones or intermediate and mountainous areas, with special attention being given to metropolitan areas. The system of a whole made up of loose ties among independent communities that diverge and coordinate, has been developed at various levels. Discussion on linkages between biospheres sometimes needs to go beyond national borders. One typical example is the Asian Network of Major Cities 21 (ANMC 21), launched in August 2000. Twelve capital cities and other major cities in Asia are now members of the Network, including: Bangkok, Beijing, Delhi, Hanoi, Jakarta, Kuala Lumpur, Manila, Seoul, Singapore, Taipei, Tokyo and Yangon. It was decided that the Network would continue comprehensive consultation on the sustainable management of cities, with themes ranging from: industry, transportation, tourism and social development to environmental issues such as: exhaust gas control, recycling of resources, water and sewerage, the heat island phenomenon and vertical forestation. Thus, decisions on policy frameworks and strategies are to be made by the major cities and their network, and communities in the major cities are to receive policy instructions from an entity other than central government.

The ideal design of sustainable development should be able to tolerate the mix of those multiple trends. One of the major issues is at which level should the organization be located in order to work as the interface between top-down, bottom-up and global forces. This organization is required to take the lead in policy coordination and power distribution among the different levels. Although there may be some arguments to the contrary, simply promoting decentralization or maintaining and strengthening the municipality-led execution system will not be enough to meet this requirement. In general, expectation is rising for initiatives at higher levels, including the regions, prefectures and unions of municipalities. At the strategy execution frontline, communities will be required to serve as bases for the transmission of information and as enhanced social systems with local creativity and cautious responsiveness. Paradoxically, communities will become more likely to be subjected to direct control of authorities at upper levels.

3.2 "Design" options for a sustainable community

Under this social environment, how should a community design a sustainable social system for itself? Figure 10.3 shows a typology of social designs. Environmental policy was first introduced under the rapid economic growth in the 1970s. In the subsequent period of stable growth, needs for an improved quality of living became more sophisticated, while the whole social system was required to shift toward a highly efficient mechanism with a lower metabolism. Several models exist for a sustainable society.

The first model is designed to scale-down the present high-metabolism social system (A in Figure 10.3). Some functions would be unsustainable in this model.

The second model (B) aims at reducing or eliminating some organs in the social system (e.g. policy organs) by introducing the principle of self-determination or self-responsibility. As a rule, this model places service providers and users close to each other so that an autonomous adjustment mechanism may work when users seek satisfaction from services. Where it is impossible to adopt the principle of self-determination or self-responsibility, the model may introduce a network principle to make active use of external systems. By building a "closed cycle" among related systems (e.g. government, non-profit organizations, and local

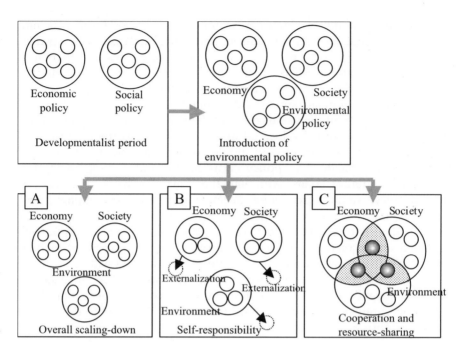

Figure 10.3 Design options for a sustainable society (*Source*: created by the author)

groups), it creates a mechanism that offsets any excess and shortage of resources within the systems without causing any spillover on the outside. However, the adoption of this model will require a new consensus, as it entails a fundamental change in the basic principle on whether to address resource problems in the social system.

The third option (C) encourages the sharing of universal resources within the social system, while redesigning the whole system into a tight-knit form with interfaces that are accessible to other systems. In particular, this model aims at enabling seamless and mutually beneficial cooperation between the separately developed systems of society and economy by integrating them through environmental policies. Thus, various entities build new conventional, subsidiary, and mutually supportive relationships in order to share their common responsibilities.

For example, in Matsumoto (Nagano Prefecture), where the "Plenty of Flowers" movement originated, the leading group of the movement (Association for a Flower-filled City) has been reorganized. With a membership of some 140 companies and associations, as well as 378 resident groups, the Association marked its 50th anniversary in 2001 by transforming itself into a single non-profit organization in order to increase the transparency of its management and further accelerate the movement.

Chiyoda Ward (Metropolis of Tokyo) also launched a Community Revitalization Program in 2001. In addition to the traditional community development activities led by block associations, a new policy was introduced to provide subsidies for larger units (newly established block association unions). The ward government does not provide any orientation for the activities. On their own initiative, residents of a broader area bring together their wisdom, picking up ideas for building their communities through a bottom-up consultation process that includes competition. The government provides logistical support for efforts undertaken by the residents.

Discussion on more advanced designs and systems than the original network structure will be required in order to improve the performance of such integrated organizations.[8]

3.3 Functions of a sustainable community

Considerable energy (resources) is needed to build a "smart community", a new design for a sustainable society. How can we define a process that enables the manifestation of the design within the framework of community development? How can individuals and society design a sustainable social system through dialogue? A hypothetical structure of this system is described below (Figure 10.4). On the left is the cycle for an actor (individual or organization). On the right is the cycle for a social system.

The "actor" cycle comprises three phases. First comes the "introduction" of an action. Individual human beings intentionally add new attempts to the existing values and experience so as to choose their own method. By trial and error, the method gradually becomes a "habit" for them and their surroundings. In this second phase, the chosen method stabilizes and is perceived as an integral part of their lifestyle, being linked to daily rhythms and resource allocation. The third phase comprises the act of "value creation (judgement)". Based on the measurement of the overall value of the new method, individual human beings judge whether the adopted method really fits in with their values, whether it represents an accidental extrinsic matter and therefore should be rejected, or whether something is missing in the method. The actor develops by repeating the cycle of those three phases to infinity.

The "society" cycle also includes three phases. First comes the "introduction" of a new social function. In this phase, a new function is intentionally added to the existing functions of the social system. Central and

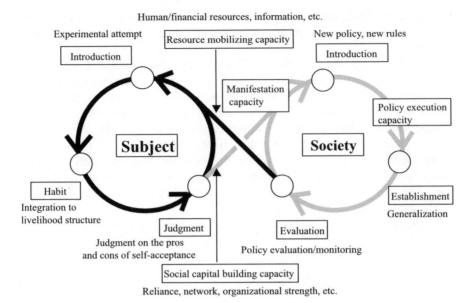

Figure 10.4 Morphogenetic process of social change and actor formation (*Source*: created by the author)

local governments as well as community groups are constantly required to introduce various new policies and rules. Important policies are introduced by the judgement of leaders or through consultation with the parties concerned.

In the next phase, the method gradually becomes, by trial and error, a "custom" for the society and its surroundings. Through social experiments and campaigns and taking into consideration the rhythm of operation, including budgets and time allocation, the method adopted is stabilized becoming something natural or rational within the society. The third phase comprises the act of "evaluation". Based on the measurement of the overall value of the new policy or rule, the society judges whether the introduced method is appropriate, whether it does or does not fit in with the society, or whether some policy elements are missing.

4. System of resource mobilizing

4.1 Resource mobilizing theory

Although the two cycles are closed and revolve separately, this section assumes that they are systematically synchronized with each other, linked like a belt-conveyor or Moebius Ring. In practice, our livelihood is not always linked directly with society. Rather, an individual or actor may intentionally elude the overwhelming influence of society in order to preserve him or herself by disengaging from it. However, what is important here is that one can re-engage at any time. An even bigger problem would be the complete disconnection of the two cycles. Indeed, the propagation of "social apathy" has become a substantial issue. The construction of a responsive social system would be necessary before discussing any vision of a sustainable society.

For the moment, let us assume that the two cycles are linked with each other (desirable interaction between the actor and the society), enabling effective transmission. In this case, the act of utilizing the merits and energy of the society to the advantage of the actor is called "resource mobilizing". We know by experience that it is easier to satisfy our desire if we ride the tide of social trends. When purchasing a new car, for example, an environmentally sound electric vehicle may bring a variety of support measures. Conversely, we need to seek and mobilize social resources if we want to realize something that is difficult to achieve on our own.

Through an analysis of various social actions and movements initiated in the United States, W. Gamson and J. McCarthy distinguished successful activities from unsuccessful ones. The findings indicate that movements may succeed or fail even if they are based on similar standpoints and perceptions. For example, when a judgement is made that an environmental activity has succeeded in District A but failed in District B, the failure is often attributed to simple "mismanagement" or "lack of effort" on the part of the leading actor. However, any movement requires another momentum, namely the "resources" existing in the environment of the actor (social system). Attention should therefore be focused on the available quantity of such resources. When no sound comes out of a radio, following this approach would lead one to first suspect that the battery is dead, and not that something is wrong with the receiver itself.

Different people have different definitions of "resources". The writer classifies resources into 14 categories: (1) human resources; (2) products; (3) financial resources; (4) wisdom/technologies; (5) services; (6) information; (7) spaces; (8) networks; (9) organizations; (10) habits; (11) rules; (12) ideas; (13) images, and (14) attachment (sense of belonging). For instance, the failure of an actor to purchase an electric vehicle may be because the actor does not know about its advantages in terms of long-term costs, easy charging, and various purchase support measures (lack of information), or because the actor has a vague notion of such advantages but considers that the application process for support measures is too cumbersome (lack of wisdom/technologies), or because the actor has the ability but not the time to apply for support (lack of habits). In any case, it is important that the actor identifies the resource gap and finds out how to close the gap, although relevant support will be required.

Attention should be focused on careful monitoring of the matching between the "design" and "system" on the one hand, and resources on the other. Although "financial resources" tend to be considered as the most versatile and convertible resources, non-marketable goods may sometimes be in short supply. "Networks" and "reliance" are among the most typical examples. For instance, any conflict between local community leaders reduces the reliance of community members and thus discourages their effective participation. Even when sufficient financial resources are available, the system does not work if the budget cannot be implemented appropriately. In such cases, necessary resources might not be additional funding but the establishment of a new habit, rule or organization.

4.2 Strategic choice approach

It may be considered that mobilization includes the following acts: (1) identifying needs; (2) recognizing; (3) abstracting/creating; (4) collecting; (5) providing; (6) using; (7) translating; (8) combining; (9) analysing/comparing; (10) projecting; (11) planning; (12) redistributing; (13) notifying; (14) confirming; (15) evaluating; (16) reconciling; (17) stabilizing; (18) determining; (19) achieving objectives, and (20) punishing transgressions.

Any shortage in a resource category has to be covered by mobilizing other resource categories. The matrix in Figure 10.5 shows the combinations of "resources" and "mobilization".

In the example, the resource categories to be mobilized are checked and linked by arrows. In practice, actual volumes should be filled in the cells whenever possible, including the number of policies, the number of individuals, the number of hours, and total value. When preparing this matrix, information concerning the procedure for resource mobilization and the weight of each resource category may be shared effectively among the parties concerned.

4.2.1 Example of description

It is assumed that the community is composed of younger generations living in relatively new residential areas. As informal exchange of information became regularized between several individuals interested in the local environment, the members began to think about useful activities for the environment. After conducting a brief survey on environmental needs, they realized that many residents complained about the lack of information in the community. They launched a formal group in order

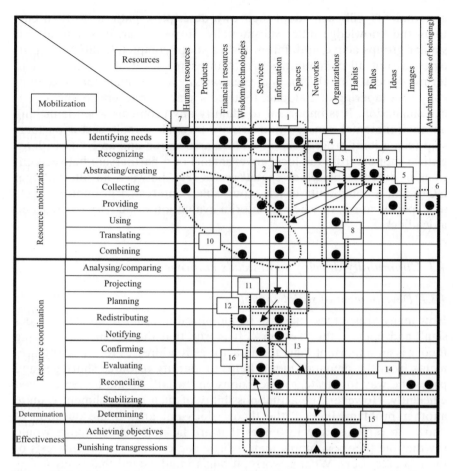

Figure 10.5 Matrix of strategic resource mobilization: example of a relatively new energetic community (*Source*: created by the author)

to provide voluntary services, e-mailing useful information on the local environment and ideas for its utilization. Local experts and students participated increasingly in the mailing list in order to discuss environmental planning and its improvement. Thus, the mailing service became a reliable component of social capital. When an environmental conflict emerged in the community, the group concentrated its discussion on the issue and informally presented suggestions to the whole community. This effectively contributed to the solution of the problem as the parties involved in the conflict took note of those ideas.

Traditional communities, community development organizations and governments will each adopt their own descriptive patterns as their customs and knowledge differ substantially. In particular, it is natural that similar groups active in different places adopt completely different strategies, as not all communities are at the same level of development in terms of social capital. This is the reason why a movement may face difficulty or a tapering in dialogue, reduced social system functioning, or social stagnation. It is important to identify any problems when launching an environmental conservation activity through a multi-sector partnership (e.g. a joint programme of a block association and an NGO with logistical support from the local government) or a partnership between entities in different places or spatial scales (e.g. between an international NGO and a community-based group solely interested in improvements in a specific area), or when its interim evaluation or activity has reached a deadlock.

4.3 Community development needs empowerment

4.3.1 Development of social capital

How many "resources" need to be mobilized in order to create the "design" and operate the "system"? It is true that the resource mobilizing theory is an important logic for an actor as it addresses resource requirements from the demand side. It helps to identify resource shortages and paves the way for strategic discussions on "what" should be replenished efficiently. However, the actual possibility of such replenishment is quite another problem. Indeed, the theory does not answer the question of "how" to replenish the resources that are lacking. Its detractors would say that the theory is simply materialistic and only serves the interests of the resource users. What is worse, it fails to understand that many actors cannot even afford to participate in the competition for resources. As noted by institutional economist J. Scott (1975), even in the face of a rare opportunity to gain maximum profit by taking up a challenge, the needy (poor peasants, for example) have to give top priority to aversive behaviour at the slightest hint of adversity (sabotage by an outsider, for example) that may result in irreparable damage.[9] Thus, only those who can afford to take on a certain level of risk may assume the challenging task of community development.

If this is the case, an enabling environment needs to be created at a higher social level in order that anyone may take up a challenge. According to Yuji Miyanishi (1986), community development means the development of a foundation to accommodate the activities of various actors, which entails the "empowerment" of the community.[10] E. Ostrom (1990) defined inherent resources in society as "social capital".[11] She argued for the necessity to question how many resources exist in the social system, whether they are available or not, and to what extent the conditionality (i.e. institution) has been developed to ensure their equitable use.[12] Only then can we start discussing fundamental questions such as: How should we repair the social system when it does not meet the conditions needed to provide a level playing field for a wide range of participants? Or should we change the status quo since few if any resources exist within the social system? Or should the actor give up trying to get the resources that are missing and seek alternatives?

4.3.2 Capacity building

At this stage, the concept of "social capacities" becomes necessary. R. Putnam focused on the "manifestation capacity" and "policy execution capacity" of an organization. At the planning level, "manifestation capacity" may be classified into the following seven categories, the capacity to: (1) identify issues; (2) find potentials; (3) understand the structure of a problem; (4) project; (5) propose plans; (6) coordinate between alternatives, and (7) adjust. Likewise, the "policy execution capacity" consists of the following seven elements, the capacity to: (1) involve; (2) communicate and explain; (3) deliberate; (4) decide (agree/share); (5) distribute fairly; (6) enforce, and (7) arbitrate. It is necessary to establish correspondence between these capacities and the theory of resource mobilizing. Building these social capacities is also an essential element of community development. In this sense, a strategy for effective capacity building is required.

5. Feedback system: Enabling conditions for social transition

So far in this chapter, the "design", "system", and "resources" of a sustainable community have been discussed. Simultaneous implementation of the approaches presented in the previous sections should lead to the creation of a sustainable community. However, how can we strengthen the orientation toward a sustainable social system even if we understand its image? Some communities will be able to get on track to reach that goal, while others will not. Why?

Magoro Maruyama (1963) proposed a "deviation-amplifying system". An actor does not only adjust itself to the environment for self-sustenance purposes. There also exists a system of positive deviational self-creation and self-generation, making use of information that will be important for the next generation. Maruyama explained this system through the function of positive feedback. The result of efforts by individ-

ual actors largely depends on whether the feedback circuit is positive
or negative, that is, whether the change is welcomed and supported. The
behaviour oriented toward the stability of present "form" and "mechanism" was called "morphostasis", whereas the action oriented toward
self-creation was called "morphogenesis". Taking into account the "positive" and "negative" feedback processes, as well as game theory, theoretical sociologist W. Buckley (1967)[13] developed a model to explain organization, behaviour, information, decision-making, roles, interaction,
social control, and institution.

According to this systematic approach (see the vertical axis of Figure
10.1), the achievement of a sustainable community requires a morphogenetic process that boosts the transformation of the community. In this
sense, we should design a circuit that prevents the ambiguity and wasteful
habits of individuals from seeping into the social system, thus creating a
new social system that coordinates relations between individual human
beings and the natural environment. Rather than concentrating entirely
on the adjustments needed to prevent the train from running off the existing track, a community is now required to detect the imminent dead-
end and give an early warning so that it may launch a process of social
transition that will focus attention on well-informed judgement in search
of a new track running in another direction.

6. Principles of a morphogenetic process

As shown by Figure 10.4, each actor and society has a growth cycle comprised of three phases, which follow on from each other indefinitely. This
cycle apparently corresponds to the Plan–Do–See cycle, one of the planning fundamentals. To conclude this chapter, this section describes the
four principles needed to ensure a smooth continuation of the cycle: (1)
conviviality; (2) strategic choice; (3) focus on policy execution capacity,
and (4) continuing confirmation of identity.

6.1 Conviviality

For both the actor and the society, the primary engine of the cycle is "introduction", which represents the first step in a new direction. This first
step is not necessarily taken with confidence that the introduction will
be successful. Rather, it may be the result of numerous half steps in various directions in search of potential development. The process of introducing a new idea tends to be curtailed on the grounds that it is not
cost-effective. Therefore, the potential of evolution and performance of
the social system largely depends on whether it has the capacity to cope
with the cost of "introduction" (human resource development, opera-

tional costs, and specialized section within the planning department). Putnam measured this "manifestation capacity" quantitatively in terms of policy announcements and regulations.[14]

However, we humans are neither sophisticated radars nor search engines. Our performance as human beings is guided by an insatiable hunger for novelties. Sociologist I. Illich (1973)[15] suggested that many of the contemporary civilization diseases are rooted in institutional stagnation and contended that a process of "conviviality" is needed to break this deadlock, so that actors may find their new identity through interaction. In a convivial process, individuals promote introductory activities, such as: having fun; stimulating curiosity; inviting people to get together; welcoming new encounters of different values; making use of informal opportunities; accumulating wisdom; preparing a strategy, and planning. It is completely different from the traditional process, in which the strong lead the way, with secondary groups following as conditions improve.

6.2 Strategic choice

The new process requires a new approach to planning. As a rule, the traditional process is implemented after its coverage, policy objective, and concrete measures have been fixed. Under ever-changing conditions, however, the process turns into an "editorial plan", focused on how to obtain maximum output by combining mobilized resources. Thus, the process depends heavily on simple coincidence. Nobel Laureate M. Eigen (1975) said that the random act of "play" fulfils an important role in the evolution of materials and organisms.[16]

In this case, intentional planning means inducing inventiveness from accidental actions in the field. It entails the act of deriving stability in form and mechanism from the infinite number of simulated combinations, as with gene sequencing or mah-jong tiles. In so doing, players compete in order to achieve the best practice, changing their objectives flexibly in accordance with the limits of resources, the available resources within themselves, and the games played with others. This type of planning may be considered as the approach of strategic choice. A community should adopt this approach to break away from the traditional method of community development, which is not based on trial and error. To ensure this, a neutral secretariat in charge of improving performance in the field is essential for a framework to promote community development, to say nothing of those actors who concentrate on their play.

6.3 Focus on policy execution capacity

It is important that "habits" in individuals and "establishment" in society follow a successful "introduction". It would be easy just to "pretend" to jump at something new, as long as there is no risk to take. Today, many of the efforts undertaken by communities, non-profit organizations, and governments are criticized for being simple "sloganeering". In following up on "manifestation", the capacity of policy execution is important, even if the policy should turn out to be unsuccessful. Capacity building, measurement and evaluation are necessary for that purpose.[17]

6.4 Continuing confirmation of identity

Finally, conscious and intentional efforts, rather than pointless actions, for sustainable community development will depend on "judgement" in the "actor" cycle and "evaluation" in the "society" cycle. At this stage, the merit of the introduced policy for the whole cycle is self-checked or evaluated objectively. For example, "positive" judgement on a sustainability policy will accelerate the movement of the cycle toward a morphogenetic process. This indicates a movement toward "introduction" with renewed recognition. Thus, positive judgements make a real difference in conditions for implementing the policy, in terms of social evaluation, cooperation of supporting entities, financing and information, among others.

This mechanism is comparable to the immune system of an organism. Malfunctioning of the system disrupts the distinction between intrinsic and extrinsic substances, eventually leading to fatal immunodeficiency disorders, such as: no reaction, overreaction (allergies), and self-destructive reaction (typically cancer and AIDS). In the process of convivial and strategic community development based on respect for diversity, extrinsic elements keep flowing into the community to be integrated in the "closed cycle". It is more important than ever to ensure that diverse actors preserve their identity by determining whether the community should accept such extrinsic elements, or modify its planning and project implementation for self-generation, or reject them altogether,[18] and by announcing the result of evaluation both inside and outside the community. Otherwise, the resource cycle might turn into a vicious cycle. Sustainable community development, based on a stable balance between the "system", "design", and "resources", hinges on these principles.

7. Conclusion

There are two meanings of "metabolism" and "adaptability" in the concept of the sustainability. So as to improve the sustainability of the physical environment of communities, it is important to pursue "design" of sustainability which can make most of the outputs more than the inputs, and to heighten the adaptability of "system" in which it can respond to such needs of urban design. The main subject of this section is the principle of this "system".

There are three important systems to enhance the adaptability to new urban design.

The first is governance system which enables to understand present conditions and the desirable urban design, and to plan the effective strategies of adaptation towards a new city, and to choose them adequately. It is essential to consider the direction of development and the role of local communities, both in central areas and suburban areas, therefore the discussion about these issues on the same table and the governance at higher level becomes important more than ever. It is important to design the system of a bottom-up process in which local communities correspond to the environmental changes, play a key role by their own accountabilities, and demonstrate independent, distributional and co-operational mechanism. The integration of local communities in the Chiyoda district of central Tokyo provides an example.

The second is the "system" of the resource mobilization. In order to build the mechanism of a bottom-up process and to operate it smoothly, a community needs to mobilize resources required for it. By mobilizing resources, it becomes possible to provide the energy for a function working, and to reorganize the community itself into tough "system". It is important to carry out the reorganizing of system itself (especially the reorganization to simple and integrated communities in order to environmental, social, and economic activity can cooperate) by the resources mobilization, but also to save the fundamental resources for community function (accumulation as social capital). In order to enable the strategic management, the method of the process design using the "resource mobilizing matrix" was shown.

The third is the "system" of feedback by which community and environment carry out an interaction between each other. It is crucial to build the feedback loop between community and environment that enables an interactive exchange of both conditions, especially one that puts a special emphasis on process execution. Designing a new "system" which accelerates that process is required to develop. It is significant to introduce and mix three rationalities in the debates of sustainability: "management rationality", which was embodied exemplarily by urban and environmental theorists such as Doxiadis and Hardin; "traditional rationality", which was evolved by some traditionalists such as Spinoza; and "communicative rationality", which was advocated typically by Habermas and his followers, such as Putnam, and from which the "knowledge" that constitutes a sustainable "system" is composed. It is communities which have the actual mechanisms to make these processes possible as part of their cultural identity that are required for future sustainable development.

Notes

1. Bruntland Report (1987) *World Commission on the Environment and Development*, p. 43.
2. International Council for Local Environmental Initiatives, 1994.
3. Motokawa, Tatsuo (1992) *Time for Elephants and Time for Rats*, Chuko Shinsho, p. 170.
4. Schmidt-Nielsen, Knut Stortebecker (1964) *Desert Animals*, OUP.
5. Ichikawa, Mitsuo (1982) *A Forest-based Hunting People: the Livelihood of the Mbuti Pygmy*, Jimbun Shoin.
6. Economist K. Boulding, who advocated the general systems theory, argued that the extent of system complexity rises from clockwork to cybernetics, open systems, animals, human beings and, finally, to social fabric. Boulding, Kenneth. 1971, "General Systems Theory – The Skeleton of Science," *Management Systems*, p. 20.
7. For typology of institutional change scenarios from the community to international levels, see Alden, Jeremy (2001) "Planning at a National Scale: A New Planning Framework for the UK," in *The Changing Institutional Landscape of Planning*, Ashgate, p. 55.
8. Soda, Osamu (2002) "Latest Developments in the United Kingdom: Accelerated Reform," *Japan Housing Conference*, November Housing Conference.
9. Scott, James C. (1976) *The Moral Economy of the Peasant: Rebellion and Subsistence in Southeast Asia*. New Haven: Yale University Press.
10. Miyanishi, Yuji (1986) "Community Development Requires Empowerment," *City Planning Review*, No. 143. The City Planning Institute of Japan.
11. Ostrom, Elinor (1990) *Governing the Commons: The Evolution of Institutions for Collective Action*. Cambridge University Press.
12. Ostrom, Elinor (1993) *Institutional Incentives and Sustainable Development: Theoretical Lenses on Public Policy*. Westview Press.
13. Buckley, Walter Frederick (1967) *Sociology and Modern Systems Theory*.
14. Putnam, Robert D (1993) *Making Democracy Work: Civic Traditions in Modern Italy*.
15. Ivan Illich (1973) *Tools for Conviviality*.
16. Eigen, Manfred and Winkler, Ruthild (1975) *Das Spiel: Naturgesetze steuern den Zufall*.
17. Putnam, Robert D. (1993) *op. cit.*
18. For a prototype of community empowerment index in Japan, see Institute of Urban and Regional Studies, Waseda University (2001) "Report on Pilot Projects for Shopping District Reactivation, FY2000", Tokyo Chamber of Commerce and Industry.

Index